pukka

Celebrating ten Pukka years

Take one herbalist, one entrepreneur, one big idea, a whole lot of inspiration and stir. The result: Pukka Herbs and millions of cups of delicious herbal tea.

"We love plants and we love people and Pukka Herbs is here to inspire healthier, happier living through the incredible power of herbs. We are passionate about improving the lives of everyone who comes into contact with us and covering the earth in organic herbs. We are blessed that at our heart we are guided by the wisdom-field of Ayurveda, the insights of traditional herbal medicine and the principles of conservation. These values help to inspire our vision of connecting people and plants and make us 'pukka'. One of the meanings of 'pukka' is 'real' and we want to keep it that way so we can make a positive difference in our world.

This book is about the shared values we live at Pukka. It's a book about how we have turned our dreams into reality. Its about how you can too."

Sebastian and Tim, Founders of Pukka Herbs

2012
Pukka celebrates 10 years of serving herbal health. Someone drinks our billionth cup of herbal tea. The Pukka family grows to 50.

2011
Pukka organic Ayurveda skincare collection arrives.

2009
'Three's' teas come out. Pukka is now distributed in 31 countries worldwide.

2008
Organic Bio Nutrient and Botanicals created. Fairtrade chais and Green teas launch. 20 million Pukka cuppas are drunk this year.

2007
The Pukka tea selection goes up from 7 to 10.

2005
We update the 'Pukka look', reach 250,000 customers and move office to house our growing team.

2003
Many visits to India later we formulate 20 Ayurvedic herbal remedies.

2002
In Spring we launch our first three herbal teas.

2001
Sebastian & Tim are inspired to spread the benefits of herbal health starting from Tim's spare bedroom.

TEN YEARS IN THE MAKING

Blessings Sebastian 31/8/12

& Tim

1 A meeting of minds

The big idea

It was the turn of the century and we, that is Sebastian Pole, enthusiastic herbalist, and Tim Westwell, creative entrepreneur, both had a burning desire to make a positive change to the world. We wanted this change to shine with excellence and so we decided to create something that symbolised everything we believed in: a cup of herbal tea. Not just any old cup of tea mind you, but an incredible cup of herbal tea that supported vital health and conservation as well as a providing a moment of bliss in the swirling pace of a crazy world.

Having tried many herbal teas, and fed up with a mouthful of dusty flavourings, we knew that something better could be done to show-off the true brilliance of herbs. Better, in the sense of offering people a more delicious cup of herbal tea whose provenance and quality was known. And better, in the sense that the blends would be authentically based on the principles of Ayurveda, the art of living wisely. We were going to make herbal teas that had depth .... and taste great.

"A person with a new idea is a crank until the idea succeeds."
Mark Twain

Seb's path to Pukka

By the time I set up Pukka with Tim I had already been fascinated with plants, conservation and herbalism for years. I had spent my early twenties after getting my degree in Hindi and Religious studies wandering around India, learning yoga, becoming a vegetarian, sleeping under the stars and walking for weeks in the Himalayas. I was searching hard to find what my life was all about. These were wild and formative times. After a couple of U-turns I realised that I wanted to spend my life working with people and plants. The dream of becoming an Ayurvedic practitioner and herbalist became a huge inspiration in my life.

So I went back to college to study Ayurvedic, Chinese and Western herbal medicine. I fell head-over-heels in love with the poetry of traditional medicine and was blessed to have some great teachers who nurtured this kernel of passion. Traditional medicine is so person-centred with such deep insights into the workings of nature that I was completely inspired. And the plants just lit me up. Throughout much of my perennial studentship I had also been working as a gardener on Hambledon Herb's organic herb farm in Somerset. I lived in a caravan for two years whilst learning the essential skills of growing herbs, composting and staying warm with spicy herbal teas in the winter. It was a great time in my life. For five years before Pukka started I had been teaching yoga classes every week as well as running my herbal practice. By now, I was thoroughly unemployable and wanted to work for myself; running a clinic was great but I knew that I couldn't reach as many people as I wanted to from my one-to-one clinical practice alone.

I wanted to really promote the wonders of herbs and I knew that a great way to achieve this would be to set up a business that would champion herbs. I wanted to create a business centred around mutual benefit, one where everyone who came into contact with it would profit: sensually, healthily, ecologically, socially and financially. I wanted to reach millions of people, so that their lives could be touched with the wonders of herbs. I thought that at the very least if they drank something delicious, or used a herb that helped their health, they would feel more positive about herbal medicine. Perhaps they would even choose to include herbs in their lives more and we could create millions of herb champions?

There was another selfish motivation. Whilst I was running my clinic I found it very difficult to get good quality Ayurvedic herbs and there was certainly no organically certified supply. Given my eclectic background of speaking Hindi, organic farming and Ayurvedic herbs it seemed like an obvious idea to try and set something up in India. So I began making speculative trips around India to farms in the middle of nowhere looking for herbs. I assumed that it would be like the UK where herbs grow as part of hedgerows and the wild habitats on farms, and I was right. On the farms were lots of herbs, such as, amla, gotu kola and tulsi. They rarely grew in any ordered way, needed organic certification and there were no drying facilities, but the herbs were there. It was the start of a long and inspiring journey.

It was important to me to be able to provide a high quality supply of herbs for my patients and other practitioners, but I also wanted to offer excellent solutions to the wider public too. My one consistent principle was that everything our new company would offer would have medicinal quality. This way we could serve both the practitioners: professionals who have dedicated their lives to helping people heal and members of the public. Our commitment to quality, in ensuring that we only sell what a professional herbalist would be willing to dispense, would be at the centre of everything we would do.

"I wanted to reach millions of people, so that their lives could be touched with the wonders of herbs."

However, there was one slight irony. I was strictly 'anti-business' and saw the commercial world as valueless and cut-throat. I had very little understanding of what it entailed to run a business but being a complete idealist, blessed with self-belief and a bushel of enthusiasm, I thought it would work. However, I needed help with the physical manifestation of the business form and that is where Mr Timothy Westwell came into my life.

Tim's path to Pukka

I was born and bred in Blackburn, Lancashire, the town of '4,000 holes' made famous by the Beatles. And I fell down more than my fair share when growing up. Luckily my persistent nature prevailed and put me in good stead to get through some tough times; from failing grammar school exams to being the fattest kid in school, life was mainly about survival. As a 'mature' teenager, I started to take responsibility for my life and made a concerted effort to lose weight. I did it, and for the first time realised that I could actually make things happen for myself and see a positive result.

Towards the end of school I thought about becoming an architect. I loved the idea of being creative and designing something that would last. However, things didn't work out as I had dreamed largely because of the normal teenage distractions of wine, women and song. I flunked my A levels and scraped my way onto an HND course to study Construction Management at Liverpool Poly. It wasn't really architecture but at least, some creativity came into it, I had escaped my home town and was living independently. In Liverpool I had a great social time but never really focused on my academic course. However, towards the end I had a light bulb moment when I started the marketing module and was inspired by Maslow's hierarchy of needs. It showed me a new way of understanding life. Suddenly, things were starting to feel as if they were clicking into place giving me a new sense of what life was about.

I came out into the working world in the eighties, the time of our last recession. For once in my life I took the government advice to 'get on your bike' and find a job. So I went south. The choice was London or Bristol. Bristol won courtesy of a space on the floor of a couple of my Blackburn chums who had found their way there. I went through a succession of sales jobs from selling office

equipment, to computer software, then to systems and business change consultancy. A great career unfolded and I learned loads but there was a twitch inside that this wasn't quite what I wanted. It kept prodding me - "so what's this all about? Are you really enjoying this? What value does it bring?". I had to stop, take space and consider what I really was about and how I could align my inside and outside world....

It took me a long time to 'escape' this soulless, political and frustrating business environment, but I needed time to discover what I really wanted to do. In making this life changing decision, it became apparent to me that I had been unhappy for some time. Yes I was earning a great salary, but my soul was dying. I knew I needed to take time off and invest in me. What did I really want to do on this life journey?

"I had to stop, take space and consider what I really was about and how I could align my inside and outside world."

During my time out, I got connected to one book in particular that really helped me understand and realise my true passions and values: 'Working from the Heart' by Liz Simpson. It provided some clear guiding points to realise that I could actually 'be me' 7 days a week, not just at weekends. I would recommend it to anyone. I started to look at lots of possibilities including environmental studies or working in a more sustainable business.

My personal health had taken a set back a few years earlier when I was told I would have to take painkillers for the rest of my life. That didn't sound very appealing to me, so I found a natural way to manage the pain with herbs, nutrition and therapy. The positive benefits this had on my health encouraged me to discover a more natural way of living. My whole life was changing and I wanted to find a project that could help me express my new-found passion for natural health. I decided to place an ad in the classifieds section of 'Venue', Bristol's cultural what's on magazine. I thought people who read it would be into a similar vibe. Perhaps there was someone with a creative idea that may need my help? The ad ran for two weeks and I got a grand total of one response, from a guy called Sebastian Pole.

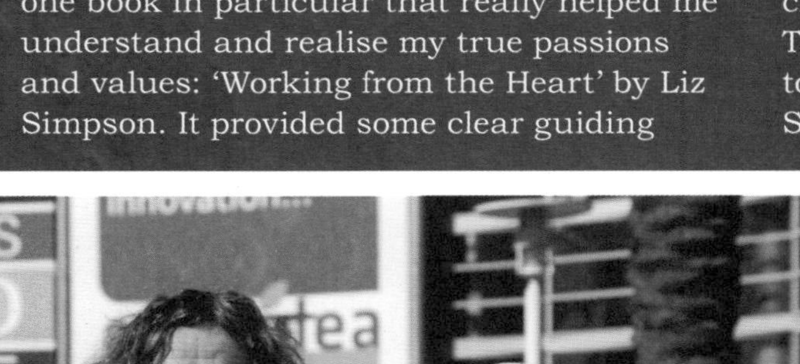

The birth of Pukka

We met through an advert Tim had put in Bristol's monthly 'Venue' magazine and bonded straight away. It was business-dating at its best. We had completely different skill sets but a common thread of shared values and mutual trust. We started to spend hours in Tim's flat in Bristol plotting our path ahead. We worked evenings and weekends trying to get our vision clarified into a brand. We had never created a brand before, but we knew we wanted to start a business with Ayurveda at its heart and spread the word of herbal health. We were going to leap off the edge and offer what we had. We knew it was good, that it had integrity, passion and essence. We just needed a name.

A name carries such importance. It's the word that symbolises who you are. After some hilarious non-starters (such as Holy Cow!) we came up with the name 'Pukka Herbs'. We loved it on so many levels; in Hindi pukka means 'real, authentic or genuine', and this was the type of herb company we wanted to create. It also means more colloquially 'ripe, juicy, tasty and delicious'. 'Pukka' symbolised all we stood for in life. Being 'pukka' was our aspiration. It also has a positive sound to it and in the UK is a slang word meaning 'great quality', as in 'it's absolutely pukka, mate'. Unfortunately it means your 'bum' in Singhalese, but let's not worry about that. It also caught the eye of a certain 'pie' company. They actually wanted us to withdraw our name. Luckily Tim knew a great lawyer whose legal skills ensured that we were able to christen our fledgling business, 'Pukka Herbs'.

So, Pukka was as much a business as a platform for promoting the wonders of herbal medicine, creating delicious teas, effective herbal solutions and covering as much of the earth's surface with organic herbs as possible. It was both passion and politics.

"Business Organics"

Offers sales and marketing expertise to help your business grow wealthy in a sustainable and healthy way

Utilise a 15 year track record to take a healthcheck or alternative view if you wish to.

- start-up selling or step-up existing sales to the next level
- develop documents outlining your sales strategy, plans and tactics

To discuss more and understand how we might optimise your organic growth call Tim Westwell

0117 974 4811

Expertise available hourly or part-time

"Dreams are meant for living."

"Ultimately, deep ecological awareness is spiritual awareness."
Fritjof Capra

Our Vision

Healthy people, healthy planet

We know Pukka has to have a positive impact. Not just on people but on the planet too. No self-righteous thinking here, just plain common sense. And a good dose of idealistic enthusiasm.
Its pretty simple. Our health and the health of our planet are intimately linked. Understanding this relationship is what ecology is all about and we wanted Pukka to be a living expression of this healthy relationship. Just as we have to be aware of our own needs to be truly healthy, so we have to be aware of the environment's needs to be truly sustainable. Ayurveda would call this 'living within our means'. Easier said than done, but by observing how our actions impact us as well as the environment, then we stand a better chance of finding a healthy balance.

By studying how Ayurveda understands nature, the life-force (*prana*) and the process of being healthy we can create a profound spiritual awareness of our own health ecology. We take this 'ecology of health' seriously at Pukka and literally go to the edge of the world to make this happen: from visiting farmers to understand our impact, to monitoring our carbon emissions, to supporting organic farming, to engaging in tree planting projects, to bringing our customers the best that herbs have to offer.

Just as we are a part of the ecosystem, it is also a part of us, and this unity reflects the interconnected nature of all life. We have to know the characteristics and needs of each component to understand the whole. This is the simple but profound idea that inspires us in creating a Pukka ecology that understands that what affects us, affects you. We know that the choices we make within Pukka have a big impact on our staff, suppliers, customers and the planet. And so we intend to make those choices wisely, with awareness.

Our Vision

Conservation through commerce

Realising that we could contribute to conservation through commerce was a lightbulb moment for us. It seemed clear that if nature has more value 'alive' than 'dead' then we will protect her. And rather idealistically, from our spare bedroom in Bristol, this is what we wanted to do with Pukka.

Through making organic Ayurvedic products we could bring some value to these threatened forests. By paying farmers and collectors above market prices for the herbs we use at Pukka we could help protect the dwindling eco-systems. By making organic farming practices worth more than soil-destroying conventional methods we could incentivise conservation of the earth. Commerce could lead to conservation.

"In the end we will conserve only what we love. We will love only what we understand. We will understand only what we are taught."
Baba Dioum, environmentalist

We saw a society in need of a deeper connection with nature and communities seeking a more natural life. It was something we craved ourselves and realised that it was something many other people wanted too. Our dream was to create a *pukka*-quality company that would bring people and plants together. Our passion for health, for nature and for the wisdom of Ayurveda was the catalyst for founding Pukka Herbs and it still is today.

"Concerted action by our manufacturers, retailers and consumers to source, sell and consume only herbal remedies which are sustainably-grown could have a significant impact on the future of nature's pharmacopeia" Martin Harper, Plantlife International.

Our Vision

Connecting people and plants

At the heart of this vision are the people and the plants that make it happen. We wanted to create circles of benevolence so that everyone Pukka touched benefited: grower, supplier, manufacturer, staff and our customers. But the real impetus for all of this was to get people using herbs. We wanted to bring a little bit of nature's garden into people's lives. We wanted to help reconnect people with their herbal heritage where plants are a central part of their health and wellbeing. We passionately believe that the smells, tastes and feelings from a cup of herbal tea alter our awareness and experience of life. Whether it's the potent phytochemicals that can influence our health or just the fact of eating something 'wild', there is a power in herbs that has led to them having been used for millennia to support health.

We want something as simple as a cup of herbal tea to be that catalyst that connects you with nature as a whole. And in connecting with nature as a whole you will connect with yourself.

"Plants are the basis of
all life, good health
and prosperity."
David Crow

Tim with the children from Chennamaiji school in Karnataka where Pukka herbs helped fund essential building works.

Our Passion

As you may have gathered, we have a passionate belief that what Pukka stands for is very important. Our Pukka creations have a serious message because we know that what we buy can have a big impact on our world, our health and even government policy. We want Pukka to have a positive impact on the political choices being made with regards to the environment, farming and to health. And we are doing this through sharing the incredible power of herbs. Pukka is a product of the times. We are a part of the global eco-health-movement where so many of us are looking for our lives to have a positive benefit. Pukka is a catalyst within this new consciousness operating as a spring-board for people to bounce off and experience the wonders of herbs. But these Pukka values are not compulsory, they are a choice and an opportunity; you can just have a cup of tea or you can delve into the heart of Ayurveda and transform your life.

We want more people to look after themselves using natural plant medicines as the first port of call before they revert to chemical pharmaceuticals. We believe in the value of the freedom to choose how we care for our health and we don't like the lack of political support for herbal medicine. We want a health system that positively supports health. In Ayurveda health is described as "the balance of body, mind and spirit; being endowed with strength, will power, zest for life and sharpness of your senses". We think this makes sense and want to support this.

We don't agree with the fundamental ideology of our modern health system that has become an illness-driven system. Whilst some of its medical discoveries are remarkable, even life-saving, we disagree with the emphasis on diseases rather than on health promotion. We are not happy that a system whose medicines ironically result in it being one of the highest causes of death in the world today, is lauded by government as the exclusive system of medicine. Our passion is for herbal medicine to be supported by government within a sustainable, inclusive and holistic health system.

And this leads us onto how nutrition is (and should be) at the heart of health. We need a system of agriculture that actually nourishes the health of the whole environment as well as nourishing our bodies. We want farming practices that build soil fertility rather than deplete it, that grow nutritious foods rather than manipulate it into 'franken-foods'. Our food can and should be grown without pesticides and petro-chemical based fertilizers. We want to know where our food comes from, how it is grown and what is added to it.

That is why we are going to the four corners of the world so that we know where our ingredients come from and to make sure the herbs we use are organic. To make sure that the people who grow them benefit as much as the people who use them. It is because traditional herbal medicine encompasses these values that we are putting all of our energy into promoting herbal health. And we absolutely love doing it.

Sebastian celebrating the harvest of sustainably grown Kutki (Picrorrhiza kurroa) in the Himalayas.

Organic inspiration

The organic movement is our religion and we pray at the altar of the compost heap. Organic stands for sustainability. You put back in what you take out. You regenerate the soil that feeds you and your family. You try and improve what you have been given.

The principles of organic farming are also embodied within the essence of Ayurveda. They support and nourish the health of the whole: air, sun, water, soil are all equally interdependent in the well-being of our ecosystem and society. Respect for the billions of microorganisms in the soil is as important as practising the highest animal welfare standards. Nurturing wildlife is as important as not spraying toxic chemicals. Organic farming really does stand for life. One of the best ways to tell that a farm is really organic is by the amount of 'weeds' that you can see. Weeds are an indication of the health of the ecosystem: attracting insects, raising nutrients from the deeper layers of soil. Of course weeds are often healthy herbs themselves, just think of dandelions! Organic is about diversity, difference and depth.

Everything we make is certified 100% organic. The Soil Association (the UK's leading organic charity), says 'Organic farming and food systems are holistic, and are produced to work with nature rather than to rely on oil-based inputs such as fertilisers. Consumers who purchase organic products are not just buying food which has not been covered in pesticides (the average apple may be sprayed up to sixteen times with as many as 30 different pesticides), they are supporting a system that has the highest welfare standards for animals, bans routine use of antibiotics and increases wildlife on farms.'

Why go organic?

There are so many reasons to go organic but one of the most compelling is the impact that organic practices can have on climate change. As we know, the last two-hundred and fifty years have seen incremental increases in atmospheric carbon accumulation leading to the 'greenhouse gas' effect and global warming. Experts agree that if this trend continues then the rise in global temperatures could have catastrophic effects on the stability of the environment and negatively impact the livelihoods of millions of people. We are very near the 'tipping-point' of 450ppm carbon (280ppm in 1700, 392ppm in 2011) but there is a solution and it requires immediate action. The Climate Change Act has committed the UK to a 34% cut in emissions by 2030. To come anywhere near meeting this target we must, amongst other things, make fundamental and immediate changes to the way we produce our food.

Organic farming continues to offer the best practical model for reducing emissions because it stores significantly higher levels of carbon in the soil and is less dependent on carbon emitting machinery and oil-based fertilisers and pesticides. A major report on the carbon sequestering ability of soil, published by the Soil Association just before the Copenhagen summit in 2011, showed that converting conventionally farmed land in the UK to organic farming could take 3.2 million tonnes of carbon per year out of the atmosphere and store it in the soil. This would reduce the UK farming's global warming emissions by a massive 23%. If the Government is serious about tackling climate change then it needs to get serious about supporting organic agriculture. Agri-business as usual is simply not an option.

We have to live according to our beliefs and if we want a cleaner, healthier and more sustainable world we need to buy organic. As the great Mahatma Ghandi said "be the change you want to see in the world", and in this case the opportunity for change comes back to consumer choice and people power. At Pukka we want to exercise this opportunity to the maximum and be a part of that change. It's why everything we grow and sell is certified organic. We are 100% organic, which in our mind, is 100% pukka.

"Organic farming is the natural solution. It nourishes the soil, increases biodiversity, creates more energy than it uses and organic food tastes delicious."

Sebastian Pole

Shri Dhanvantari,
The Lord of Ayurveda

Ayurvedic inspiration

Ayurveda is amazing. It's India's ancient system of health and yet so much more. It's a way to live and a way to understand and transform your life. It is often translated as 'knowledge of life' and encompasses the idea of how to live wisely. In particular it is the knowledge of how to live according to your *unique* and *individual* constitutional make up that puts the choices of how you exist firmly in your court. A description given in the *Charaka Samhita*, an early Ayurvedic text, written in about 100BCE says:

"It is called ayurveda because it tells us what foods, herbs and activities enhance the quality of life, and which ones don't."

It really is that simple. The essence of Ayurveda fits perfectly with what we want to share at Pukka; we want to plant, grow, harvest and blend herbal teas and remedies to enhance life.

At its core, Ayurveda teaches respect for nature and an appreciation of life, by showing how we can empower ourselves as individuals. It understands that our individual health cannot be considered as separate from anyone else's; from our family, work, society and planet's health. It describes in intricate detail how you can fulfil your potential. What could be a better guide for us to have at the centre of Pukka's values?

The wisdom of Ayurveda is an anytime, anywhere, anyone sort of wisdom. It is expressed as a way of life that flows with the changes of the seasons, weather, time and place. It teaches dietary and behavioural adjustments that can be adopted as you mature from childhood through to adulthood and into old age. It gives advice on how to prevent illness as one season becomes another, and specific recommendations on how to adjust your daily habits. This way of wholesome living prescribes a routine for all the different climates and geographical regions of the world. At the root of Ayurveda is its focus on the uniqueness of each individual. As such it is a universal system applicable to every individual living in any part of the world. It is personal medicine at its best. It really is amazing.

If you want to find out more why not have a look at Sebastian's book? Its called 'A Pukka Life' and is really a Pukka Tea in a book about life, health and vitality.

2
Living the dream

Make tea not war

A simple cup of herbal tea seems like such a good way to express our values, introduce people to herbs to promote conservation. A cup of Pukka tea could even be the first step on the ladder to finding out more about Ayurveda and, perhaps, even yourself. It is a lot to hope for from a cup of tea but at the very least it is a guaranteed good moment in the day. If we can turn our day into a succession of good moments we can really have a good day. And a good day can turn into a good week. And a good week into a good month. Hopefully even a good life.

By the time the founders of Pukka met in 2001, Sebastian had been experimenting with making herbal teas for quite a few years. In fact, our Pukka Love tea, grew out of a gift he made for his girlfriend before Pukka was even born; a little bag of hand-blended herbal tea of flowers including roses, lavender and limeflower. The prototype 'Love' tea must have cast its spell as a couple of years later they were husband and wife.

Blending herbs is both an art and a science. It's a bit like cooking; you have to know what ingredients work together in what proportions to create a culinary delight. It's the same with a herbal tea, but as well as searching for delicious flavours, you are also combining herbs for their specific health benefits.

So, our first teas had a lot to live up to. And in our idealistic enthusiasm the first teas we launched were authentically Ayurvedic. Blended to suit the Ayurvedic constitutions of *vata*, *pitta* and *kapha*, they still remain today as Relax, Refresh and Revitalise. These first three teas arrived in the early Spring of 2002 at the UK's largest natural foods trade show in London and all of a sudden Pukka was real.

Inspired by natural beauty

The majesty and wonder that nature inspires in us is within the very essence of Pukka. Her elegance and beauty is something that we have wanted to emulate in everything that we do: this idea has been at the heart of our packaging designs and is a question we ask ourselves; how can we use the quality of beauty to make the world a better place? Ayurveda identifies beauty as one of the most important parts of healthy living. Seeing beauty makes us feel good, just as feeling beautiful makes us see the world in a better light.

And in order to manifest these ideas we have been very fortunate to work with Nina and David at The Space who have managed to take our vision and put it into beautiful designs. They have been able to blend our vision of connecting people and plants so that our packaging looks as delicious as the herbs inside. Beauty really does taste good.

pukka
love
pukka
lemongrass
& ginger
organic herbal tea
delightfully fresh & uplifting
20 tea sachets

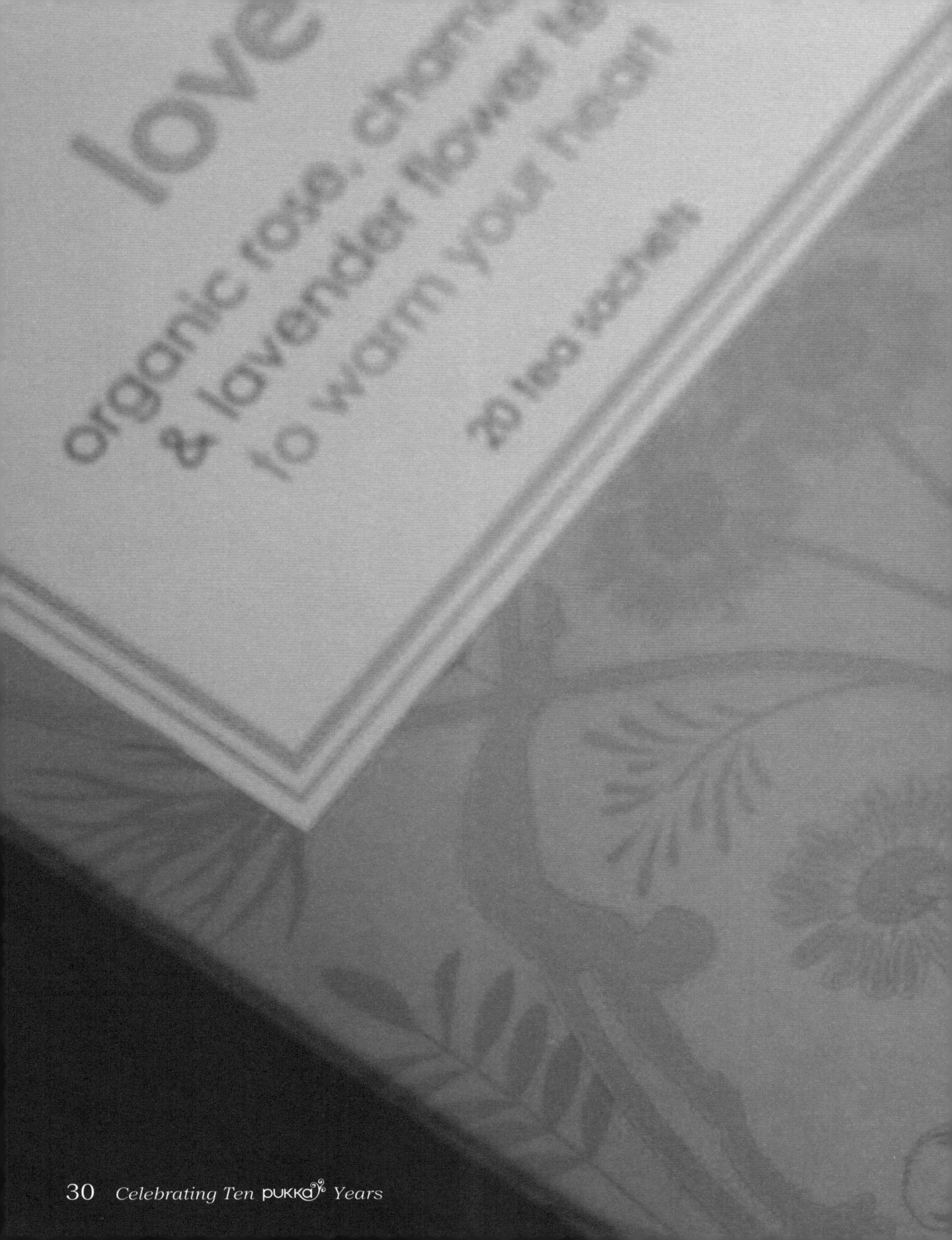

Taking teas to market

Once we had our first teas we needed to get out on the road and inspire people to start stocking them. We wanted to get everyone tasting the new Pukka teas. The market was saturated and there were already more types of tea than there were loaves of bread so we had to do something special. We went to our local Health food shop wholesaler, Essential Trading in Bristol and met the buyer Steve Penny who proved that buyers can be human. Steve gave us lots of advice, support and guidance to help us on our way. We started to discover that the natural products industry was an empowering community where people were passionate about working together to promote natural and organic foods. As we secured a few listings our confidence grew and we set our heights on the crème de la crème of organic independent health stores, which back in 2002, was Fresh and Wild in London. Aylie Cooke was head of 'grocery' and in charge of tea buying. She never minced her words and we respected her opinion of what we needed to do to make Pukka a success. We also became good friends, and knowing that we had no money, she let us sleep on her floor when we went to do in-store tea tastings in London. It was a genuinely fun time. The flavours were getting out there and a little ripple of a 'buzz' was spreading.

It wasn't long before we won our first award for 'Best New Organic Product' at the Brighton Natural Products show 2002 and the Pukka word started to spread.

We now have so many friends, colleagues and people we admire in the health food trade. The independent shops are a massive inspiration to us. They are pioneering warriors fighting for the cause of natural heath. As we spread and grow they will always remain our heartland.

Growing the business:

Introduction of Ayurvedic remedies, Organic Bio Nutrients and Botanicals

As our teas started to flourish we were also building our range of Ayurvedic remedies, Organic Bio Nutrients and Botanicals. We wanted to create something incredible which involved finding the right ingredients that are concentrated in the right way. So we found sources from every natural habitat: freshwater, seawater, desert, forest, jungle, mountains and farm land. We found specialist manufacturers who make 'whole' herb extracts including freeze dried juices, supercritical extracts, galenic fluid extracts, tinctures and water extracts to ensure that you are getting the right form of herb for the job. These are some of the most vibrant and concentrated forms of nature you can find. They are literally awesome.

In 2003 we launched our Ayurvedic herbal remedies for practitioners and the wholefood trade. As the demand to educate became even greater we developed our first training manual for in-store learning. But we couldn't do it alone and we needed help. It came in the form of Adele Lewis, who loved the health food trade and Ayurveda even more. Adele's passion and understanding was infectious. Over the four years she was with us she gained the highest accolade in the industry. At the Natural Products show, she won the prize as best sales person of the year, voted for by the health food retailers of the UK. We were on our way...

pukka
organic
virgin coconut oil

UK growth

Simultaneously with our glimmer of growth in the independent health food trade we knew that in order to fulfil our vision of spreading herbal health we had to reach far and wide. The supermarkets with their huge distribution and customer base offered a unique platform for this. In order for the world to become more 'organic' we had to try and make our teas accessible to everyone, not just the converted. More Pukka teas meant we were creating the need for more organic land. It also meant we were able to plough our profits from these tea sales into training and education about Ayurveda, herbal health and the organic movement back into our health food shop heartland,whilst providing a springboard to start our international business.

Pukka around the world

Meanwhile things were beginning to stir abroad. In 2005 we had a call from Denmark: from Diana and Henrik, yoga teacher and professional ice-hockey player. We got to know them and found they too shared our same values and passion, in fact they were bursting with it. So in the Autumn of that year, from their spare bedroom in Copenhagen, Pukka was launched in Denmark.

In the next couple of years other Scandinavian countries joined us on the Pukka bus. Scandinavia was becoming our Euro heartland as we were joined by Marianne from Norway and Marie from Sweden, both passionate health-conscious eco-warriors.

Next came Italy and Israel with Oliver and Zipi, who both had an inspiring vision for taking Ayurveda to their customers. In 2005, our first speculative trade show in the Far East in Japan resulted in finding a distributor there. Becoming an inspiring and effective leader in herbal health became the vision and Tim completed his first global circumambulation when he landed in the US and decided to start the long journey of selling our teas into the Union.

From our humble origins in south-west England, we export to over thirty countries and our international business today represents a significant part of our overall business.

"We have worked with Tim, Sebastian and the Pukka team for many years and are passionate about Pukka. Behind each product, there is always a great respect for the earth and the people who have grown and supplied the ingredients."
Diana and Henrik Carey, NatureSource, Denmark

"We decided to work with Pukka because from the very first meeting, we met authentic people, who have the passion to create healthy products at the highest level and practiced sustainability in the most transparent way, we've ever seen."
Peter Schneider
Pukka Herbs, Germany

"Pukka pioneered the growing of certified organic Ayurvedic herbs from India. We deeply appreciate their commitment to growing plants sustainably and practicing fair trade principles with the farmers. The Ayurvedic community in America has benefited greatly from their efforts and we love working with them."
Scott G. Cote and Kevin J. Casey, Co-Founders of Banyan Botanicals

"When I first met Pukka and Tim in 2007 at the Bio-Fach fair in Nuremberg, I intuitively felt a kinship and a wish to develop a business relationship. I strongly believe that through their work a little seed of awareness is awakened in many souls, even through drinking a nice cup of tea... one never knows."
Oliver Graf, President
Inner Life, Ananda Assisi Soc. Coop, Italy

"Knowing that each herb in every tea has a purpose is important."
Marianne Willumsen, Pukka's Norwegian Distributor

"We just love the positive feedback we get that Pukka really helps our customers to live their life fully."
Zipi & Rafi Mosseri
Our distributor in Israel

Despite the well-designed branding and professionalism of the products there is still something home-cooked and intimate about Pukka."
Al Overton,
Buyer at Planet Organic store, London

Pukka reaches the shores of California, USA

World map of Pukka sales

Australia
Baltic States
Belgium &
The Netherlands
Canada
Czech Republic
Denmark
Finland
France
Germany
Greece & Cyprus
Hong Kong
Hungary
Iceland
Israel
Italy
Japan
Korea (South)
Lithuania
Malta
New Zealand
Norway
Portugal
Russia
Slovenia
South Africa
Spain
Sweden
Switzerland
Taiwan
United Arab Emirates

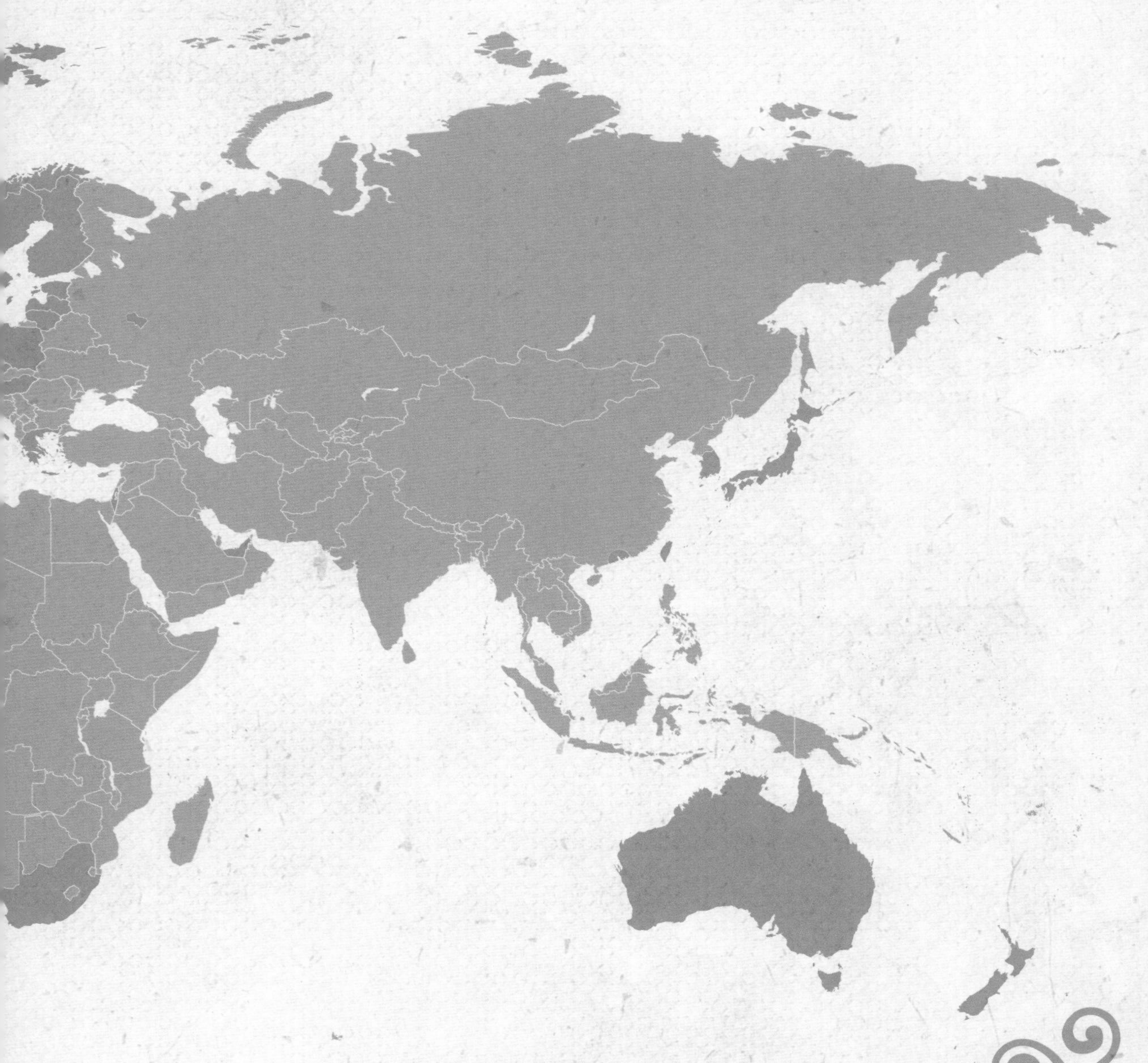

pukka

Blending it all together

Respect for our partners who play a part in Pukka's success has been a cornerstone of our ethos from the beginning. By respect we mean something much more than good manners. It's an appreciation that all our stakeholders play a vital role in helping us bring people and plants together. Each of them is a vital link in the Pukka family, from farmers to manufacturers to tea drinkers, and we respect that they also have their own unique needs that need to be considered whenever we speak with them.

It's not always the most glamorous part of our business but how we manufacture our teabags, capsules, freeze-dried extracts, supercritical extracts and juices is essential to the quality we bring to our products. And this means teaming up with experts in their field. Some of the equipment is rather incredible too. To give you just one example, our tea-bagging machine can precisely drop 2g of herbs in a tea bag, stitch a string in the top and wrap it in a protective envelope at rapid speed. It actually produces 12,000 tea bags an hour! True engineering brilliance.

Introducing Pukka Ayurveda Skincare

Because beauty is such an important part of health it seemed a natural progression for us to develop a skincare collection.

Creating beautiful products with beautiful ingredients to help celebrate natural beauty is what we are all about. Just as a natural sense of beauty is a result of a healthy lifestyle, beauty itself also stimulates good health. When we looked at the long list of unpronounceable synthetic ingredients present in most skincare ranges it became clear that we had to create our Pukka Ayurveda skincare collection. We knew we had the passion and the skills; Sebastian had a decade of clinical experience as a herbalist specialising in traditional dermatology. He also had a long-running obsession with certain essential oils. We also had Samantha Entwisle, our aromatherapist and herbalist, who is an expert on essential oils and Amanda Heron, our sales manager with deep experience in the natural beauty-care field. So having made the decision to create our own skincare range we set out to create something special.

Ayurveda has always looked after the health of the skin and has incredible insights into how to protect and rejuvenate it. Such simple ideas that are at the core of all Ayurvedic wellness, such as helping the metabolism, cleansing and rejuvenating, can be applied to the principles of any skincare collection. All of the Pukka Ayurveda creations were blended with these key principles in mind. And to feed this creativity, in true Pukka style, we searched far and wide to source some truly incredible ingredients grown by remarkable growers. Amongst other gardens of Eden, we found delicious shea butter in Ghanaian forests, sensual rose buds in the Bulgarian valleys and exquisite neroli blossoms in Egyptian orchards.

At the heart of our skincare are the incredibly concentrated 'supercritical' extracts made from some of Ayurveda's most revered skin-healing herbs. These extracts are known as 'supercritical' because they extract the fat-soluble portions of plants at the 'supercritical' phase when carbon dioxide changes from a gas into a liquid. Its environmentally sound and an incredibly potent technique for getting the active compounds out of a plant. For example, we take 720Kg fresh gotu kola leaves and concentrate them down to 1Kg of supercritical extract. Supercritical extracts really are that intense and so it means that a little goes a long way.

The world is not short of skincare brands so we want to make sure that Pukka Ayurveda is making a positive difference within this huge market. It needs to be better than what is already there. Better in the sense of its impact on your health and its impact on the planet. And, along with making you look amazing, our skincare needs to smell and feel great. The Pukka rules apply: the values are grounded in traditional wisdom, the ingredients are the best, everything is certified organic, where possible the ingredients are fairtrade, the tubes are made from recycled plastic (we developed the highest percentage recycled plastic tubes to date for this collection), nothing is tested on animals, and there are no 'nasties'. Just the pure delight of organic flowers, juices and essences, inspired by natural beauty. Beautiful.

The complete range of Pukka Ayurveda organic skincare products launched in 2011

WINNER
NATURAL HEALTH
MAGAZINE
Beauty Awards
2012

pukka
AYURVEDA
NOURISHING NIGHT CREAM
ORGANIC FRANKINCENSE & AVOCADO
TOTAL REPAIR COMPLEX

pukka
AYURVEDA
RADIANCE SERUM
ORGANIC ALOE VERA & MANUKA HONEY
REVITALISING FORMULA

pukka
AYURVEDA
FIRMING FACE OIL
ORGANIC STARFLOWER,
ARNICA & ROSEHIP OIL
MATURE SKIN FORMULA

pukka
AYURVEDA
NOURISHING FACE OIL
ORGANIC ROSE OTTO,
AVOCADO & POMEGRANATE
OMEGA OIL FORMULA

pukka
AYURVEDA
FIRMING EYE CREAM
ORGANIC BLUE CHAMOMILE
& ROSEHIP OIL
ANTIOXIDANT FORMULA

3
Great growers

Great Growers

Of course, the farmers and the collectors hold a special place in our Pukka partnerships. Their deep knowledge of nature, the soil and plants is a continual source of inspiration to us. Their hard work is legendary.

The relationship we have with our growers is one of the most rewarding things about our Pukka journey. We visit them regularly to explain our needs to them and understand the challenges they face. Most of the farmers we work with just own a small piece of land and join together with their neighbours to form a group and sell to a larger producer who can handle all the pleasantries of shipping and export. Whilst we have a long-term and committed relationship with the growers, we in no way, own their crops. They are free to sell to whoever they want. However, we work hard to make sure that we are top of the list.

Mr CMN Shastry

We have worked with CMN since Pukka started. Based in South India, he is no conventional farmer, more a visionary for a better world. He sources and facilitates the growing of over 100 species of organic herbs on over 5000 acres of land. We buy such herbs as Ashwagandha, Brahmi, Ginger, Tulsi and Turmeric from him. Pukka and CMN both started in 2001, and its been very rewarding watching us both grow together; he has recently built a first-class processing facility to handle our organic Ayurvedic herbs.

Although CMN loves plants he is really interested in helping people. That is his 'thing'. Social mobility is not always easy in India and CMN specialises in helping those who want to help themselves. Along with his partners Adiga and Viji, he has built a truly inspiring business that lifts many of their employees into a much more optimistic life. It goes without saying that he is also a pioneer in the organic movement in India and a committed conservation activist.

Every year CMN and his team organise a tree-planting day during which people from the surrounding villages, including pupils of the local school, join hands to plant 108 rare and valuable species in the area around the temple and its sacred grove. The team has also dug more than 500 rain-water harvesting trenches to capture water during the monsoon and recharge the groundwater. In the last few years the villagers have started to see a real difference; some farmers who were previously only able to grow one crop in a year now have enough water in their ponds to grow a second crop during the dry season. When you live off what you can grow doubling your output makes a huge difference. Conservation works.

CMN Shastry standing in a buttress of an ancient tree in a sacred grove on his farm in South India

Vikas Bhat

If ever there was a green-fingered man inspired to protect nature, we find it in Vikas. We have worked together for the last ten years to source the world's first organically certified Chywanaprash as well as some classical Ayurvedic oils. He has devoted his passion for conservation into sourcing and growing a sustainable supply of some of Ayurveda's most endangered species. His farms are the epitome of organic farming at its best. One is even called 'Ananada', meaning the farm of bliss. And this love of nature also extends to his animals. His small dairy herd of indigenous India cattle that he keeps for making the ghee in our Chywanaprash are very well looked after. They even get Indian spiritual music played to them everyday through a slightly croaky tape player. His bees are kept to the strictest organic standards of the Soil Association that are even higher than European standards. But his piece de resistance is his love of feeding the monkeys. The grounds in his factory are also home to a troop of monkeys who know that he will feed them everyday. They are also home to a large cobra.......

Mike Brook

Mike is the epitome of an organic pioneer. He has helped Pukka source organic herbs since our inception. With a far-reaching vision and a passionate determination to develop lasting relationships with growers, Mike is a leader in the British organic herb community. Sebastian spent a few formative years living in a caravan on Mike's organic herb farm. He knows what a hard worker Mike is and what hard work it takes to grow, harvest, clean and dry herbs. Mike has developed very efficient ways of growing herbs in the UK and has inspired a steady trail of enthusiasts to learn from him.
He started by supplying herbalists from his organic herb field in Oxfordshire and has transformed his business, the Organic Herb Trading Company into supplying some of the leading organic food producers in the UK. Mike was the first person in the UK to import organic Chamomile, Ginger and Peppermint.

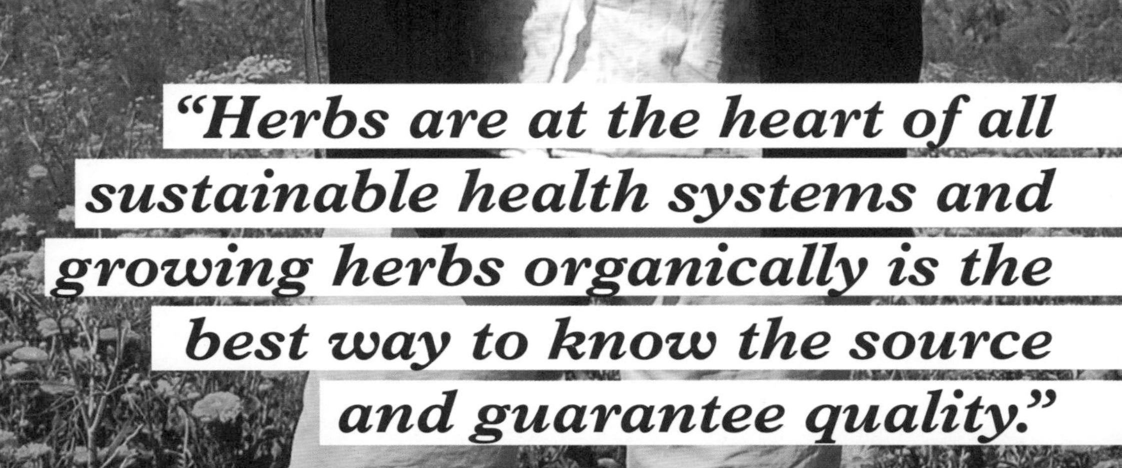

"Herbs are at the heart of all sustainable health systems and growing herbs organically is the best way to know the source and guarantee quality."

Rose growers

Our Rose growers are a wild bunch from Bulgaria. Rising before sunrise they pick roses until the sun gets too high. They work the fields early in June every year, filling bags with the fresh petals ready for our Love tea. Remarkably, it takes 1,000,000 rose buds to make one litre of rose oil that we use in our organic skincare. True steam-distilled rose oil, officially known as Rose otto, is pink gold. Our visits to the rose harvest has confirmed that roses are the symbol of love; as the air becomes filled with the exotic scent of rose blossoms, the pickers' cheeks become rosier and rosier and the laughter in the fields become louder and louder...

Bhasraj & Father

Bhasraj is a classic organic entrepreneur. Living in Northern Karnataka in South India he was brought up working on the small family farm growing cotton and chillies. After training in agriculture and organic farming, he started working as an inspector monitoring other farmers who were growing organically. When he saw how productive and beneficial it can be he persuaded his family to convert their family chilli farm to growing organic herbs. It is now an oasis of organic herbalism. As you can see his father, Devappa, looks rather proud of him.

Nagaratna

Nagaratna has been tending and growing Brahmi for us ever since we started to grow it in our special sacred Brahmi fields. She heads the 'organic' team in South India; planting, weeding, composting, fertilising, harvesting, washing and drying the brahmi and usually under the baking-hot sun. *Brahman* is the Hindu name given to the universal consciousness, Brahma is the divinity responsible for all 'creative' forces in the world and 'brahmi' literally means the 'energy or *shakti* of Brahman'. With Brahmi deriving its name from these lofty roots, it has a lot to live up to. And it does! Its mind enhancing and nervous system soothing effects are legendary. The care and attention that Nagaratna brings to tending for our crop is equally incredible.

Brahmi is a water-loving plant known for raising spiritual awareness

Ganapatti

Ganapatti is a nature lover. He lives in the forests of the southern Western Ghats. He knows so many of the plants in the forest that a walk with him is a herbalists dream. As you can see, he is a rather happy man too. His vitality exudes that of a man inspired by the verdant life-force in nature. He grows Turmeric, Ginger and Kapikacchu for us. We are sure some of his positive energy finds its way into the herbs.

World map of Pukka ingredients

pukka

4 *Vital values*

Our work ethics

What we believe and how we behave

3. Be brave
Stretch your limitations and be a compassionate warrior.
In our mission to bring the power of herbs to the lives of millions of people, we will surprise, stimulate and dare to be different.
4. Be wow!
Never stop being amazed by the wonder of nature; its beauty, its majesty, its incredible healing power.
We remain gracious in sharing our wisdom and strive to WOW each time you see, taste, smell or hear us.

Building a values based and vision-led business

We want to create a business whose *being* is reflected in its *doing*. As you know, much of our inspiration comes from the values found in Ayurveda and the larger wisdom tradition that is Indian culture. But theory and practice are two different things and somehow we needed to translate these ideals into some practical exercises so that they could empower our business.

The ancient Indian texts, known collectively as the Vedas, that embody this wisdom tradition, describe how this can be done. They say that everything in the world is upheld or supported by four principles:

- Knowing what is the essence of something.
- Knowing how this essence relates to others.
- Knowing what is the quality of this essence.
- Knowing what is the energy that maintains this essence.

The lives of truly successful people and companies are maintained by a delicate balance of these four aspects.

These concepts can be condensed into the values of truth, respect, purity and effort. In Sanskrit these are the qualities central to the practice of yoga called *dharma, ahimsa, shaucha, tapasya*.

For example, your *dharma* is your life's path, it is the truth of your life. So, to follow your *dharma* is to follow your truth. 'Pukka' is synonymous with this original Vedic word '*dharma*', meaning 'that which sustains'. Pukka and *dharma* also share a similar meaning: 'authentic' or 'true'. *Ahimsa*, means 'without-violence', and is the attitude of respect that we hold when we relate to other people. *Shaucha*, means 'purity' and relates to the standards that we apply to our actions. *Tapasya*, means 'effort' and is the energy we put into our work.

Taking these concepts of truth, respect, purity and effort as a guide to creating a strong company that is successful and fun to work for we regularly ask ourselves four questions:

- Who are we? – What is our identity and what do we stand for?
- What do we do? – How do we affect our world? What experiences do we create for our employees, suppliers, customers and our environment? How can we make it better? How can we create the most benefit with the least harm?
- What are our standards? – What guides our actions and decisions? What is our plan and how clear is our communication? How do we know when we are truly Pukka?
- How do we work? – How do we spend energy? Are we elegant and efficient in what we do? How can we get great results with the least effort?

These questions sometimes bring about some difficult answers, but they always lead to improvement and enable us to move closer to our truth. Asking these questions helps us to self-reflect and brings us a greater sense of self-awareness. Just as sometimes we pat ourselves on the back; at others we just have to laugh at our folly. Like most journeys, it's not been a smooth ride to the heart of Pukka, but it has certainly been enlightening.

With thanks to the Cranmore Foundation for helping us develop these ideas.

The importance of values

There is no doubt that having a values-based approach has resulted in loyal customers, long-lasting supplier relationships and increased demand for Pukka in markets all over the world. However, the excitement of a growing business and the logistical demands of expansion is an ever-present reminder of the importance of staying true to the values that got us this far. For us this means staying focused on our passion to be exemplary.

Responding to growth has obviously required investment in people, systems, quality control, production, working with an expanding network of farms, and working more closely with suppliers of essential services who have invested heavily in their own facilities to meet Pukka's growth. It really has been a journey of Pukka growing from a child to teenager with our staff especially having to adjust in a climate of perpetual change.

But there has been one constant. And that is that everyone at Pukka makes continual efforts to be the best. As we experience new challenges all of our partners strive to solve the problem every day, day after day. And this relentless passion for being the best whilst also living ones values highlights how essential every partner in Pukka's business ecology is to us providing the best products. Beginning with our dedicated employees, to our suppliers and farmers, our distributors and loyal shops that stock our products, everyone in that chain is part of our value system and our ability to meet the needs of a growing international market.

Pukka is grounded in our values. When we, Sebastian and Tim met there was an instant connection and we could really work together because we shared the same vision and values. We might not have always had the same approach or style but we have had the same intention. And our willingness to learn means we have developed a deeper understanding of who we are and what essence guides us to live by those values. Many people have helped us forge this path, and if we have been the two musketeers then, our Chairman Gil Williams, has been the third. His guidance has helped us immensely in anchoring our very fast growing business whilst keeping true to our values.

"Everyone at Pukka
makes continual efforts
to be the best."

5 *Incredible Herbs*

Why are herbs so remarkably effective?

"We live by a small trickle of electricity from the sun." Albert Szent-Györgyi

There are many ways to understand how herbs work. The traditional view is that they are filled with an energy, known as the life-force. One of the most wonderful and profound teachings of Ayurveda – in fact of all traditional medical systems – is this notion of this energy that pervades all existence. In Ayurveda it is known as prana.

Everything that is alive is energised by prana: the fresh vitality of a waterfall, the tingling in the hands after a clear meditation, the sweet bliss of feeling love – these are all tangible living experiences of the life-force. Within you, prana brings immunity, warmth, strength, enthusiasm and vitality. It protects you from infections, keeps skin firm, and brings a sparkle to your eyes. This life-force empowers your entire being, as it permeates every molecule of matter.

The life-force is like a current of electricity. Green plants, trees, herbs and vegetables all harvest the sun's life-force as light via the process of photosynthesis. When we consume these foods, this stored solar energy is released into our bodies. It is then transformed into the biological energy necessary for all cellular function and life.

Another way of understanding how herbs are so effective is through their phytochemicals. In order to protect themselves from bacteria, viruses, fungi and the damaging effects of the climate, plants develop certain molecules to protect themselves. These are known as secondary metabolites (in contrast to primary metabolites, such as protein, carbohydrates, fats) such as tannins, alkaloids and volatile oils. As humans and plants have evolved together over millions of years we have developed a symbiotic relationship to benefit from these protective plant molecules.

A health solution must:

- Effectively remove the causes and the symptoms of illness.
- Improve health and vitality.
- Be safe with minimal or no side effects or risk from harm.
- Be practical to use.
- Be affordable for the individual and for society.
- Be sustainable

Herbal medicine meets these criteria in abundance.

How natural is your Vitamin C?

To give you an idea of the philosophy behind how we make some of Pukka Herbs' supplements lets take a look at Vitamin C. Vitamin C is renowned for supporting the body's natural defences and is one of the most popular food supplements used today. However, most Vitamin C supplements available are industrially produced ascorbic acid, a reduced form of this naturally occurring vitamin, which comes from genetically modified and bio-synthetically manipulated corn sugar. Because the ascorbic acid is synthesised, it is far less nutritious than its natural counterpart which needs to be in a food-form that is best absorbed when taken with an array of nutritional co-factors (such as anthocyanins and bioflavonoids). Your Vitamin C supplement is probably not as natural as you think. Like much of life, Vitamin C is most effective for us in its natural form. Blended with organic fruit concentrates of acerola, amla, and bilberries Pukka's Natural Vitamin C is not a heavily processed or corn derivative, but a highly effective food supplement that supports our immunity. Due to the fact that it is totally natural, it is more slowly absorbed by our bodies and stays in our blood for up to 12 hours, far longer than most bio-synthetic forms of ascorbic acid. One daily serving delivers 250mg - or 312% - of our recommended daily Vitamin C allowance, and has a greater greener footprint than the industrially processed form since it doesn't require such high levels of energy consumption during the production process. Its really is natural Vitamin C as nature intended.

We love herbs. We are obsessed with them. Their beauty, their delicious flavours, how good they make you feel; their ability to cleanse your system as well as nourishing the strength of your body. Meet some of our favourites.

Life story of Ashwagandha

Ashwagandha (*Withania somnifera*)
Ashwagandha root is one of Ayurveda's most important herbs. Its also one of our favourites at Pukka. Its strengthening, calming and rejuvenating properties have kept us going for years! It is an annual plant that grows up to two and a half feet in height with small oval shaped green leaves and flowers that are lucid-yellow to green. Its also known as 'winter cherry' because of its characteristic 'Solanacea' small berries that are orange-red. We grow our Ashwagandha on an organic project in South India where it thrives in the light-loamy soil.

We sow the crop at the beginning of the rainy season in July and then, weather depending, harvest in January. The short, white tap roots are pulled straight out of the ground by hand and dried for a few days until they are crisp enough to snap.

Ashwagandha's reputation for promoting inner calm and core vitality makes this herb perfect for the 21st century. Ashwagandha is particularly good for those who are anxious and have difficulty sleeping. Its ability to strengthen the blood, build iron levels, keep muscles firm and increase bone density make it indispensable for slowing ageing. This herb also has an incredibly diverse effect on the key stress and hormonal triggers in the body, with an affinity for the adrenal, endocrine and nervous systems. It has a long history of use in immune disorders, hormonal imbalances, thyroid problems, blood sugar levels, chronic inflammation and infertility.
It really is perfect!

Life story of Ginger

Ginger (*Zingiber officinale*)
Ginger is one of the most commonly used spice herbs in the world. It can grow up to a metre with lush, languishing, lanceolate leaves and has a stunning red flower. Although we call it 'Ginger root' it is actually a rhizome; rhizomes are underground stems that store the plant's nutritional and medicinal properties. That is why ginger is sometimes called 'stem ginger'.

Ginger is a perennial species, but for the purpose of cultivation we treat it as an annual, planting it before the rainy season in May and harvesting six to nine months later after the flowers have appeared. Most of our ginger comes from South and East India where the heat of the sun seems to penetrate the root and make it deliciously spicy.

One of its Ayurvedic names is '*vishva-bhesaja*', meaning the universal medicine that is good for everyone and everything! But normally when using ginger, we think of the digestion, lungs and circulation. Its digestive benefits are legendary: it warms and strengthens the digestive system, and increases digestive *agni* and the secretion of digestive enzymes. It is useful for keeping the digestive system healthy by preventing nausea (morning, post-operative and travel sickness), flatulence, griping pains and sluggish digestion. Its warming properties help to get mucus off the chest and to clear a cough. A hot brew of ginger is really one of the best things to quickly warm the extremities and keep a winter chill at bay.

Life story of Tulsi

Tulsi (*Ocimum tenuiflorum syn. sanctum*)
Also known as holy basil, there are two main types of cultivated tulsi: rama tulsi, with green stems and leaves, and krishna tulsi, with greenish-purple stems and leaves. Both varieties grow up to about half a metre with the most delicate white flowers. We also grow a type of lemon tulsi called *Ocimum americanum*.

Tulsi is a survivor and grows well in a wide range of climates and soil conditions. We grow it with farmers in South India where it is fairly humid, with decent rainfall but not too dry nor damp.

Tulsi is a favourite of farmers as it is a prolific crop to grow. It is normally ready for the first harvest three months after planting when the plant is in full bloom. It can then be harvested again at one and a half month intervals for up to two years. It is considered to be such a sacred plant that some farmers we work with will not let us wear shoes in the fields where it is grown.

Tulsi is a member of the mint family and has similar fragrantly aromatic properties. This 'lightness' is thought to confer its ability to lift the spirits, ease depression and help anxiety. Its spicy-dispersing action makes it very useful for protecting from seasonal colds and fevers, as it reduces damp kapha and cold vata. It is also very useful for tension headaches, allergic irritation, nervous digestion and the physical ache associated with 'flu and colds'. So, overall, Tulsi, is a potent rejuvenating plant that builds immunity and raises your spirits.
That's why its Holy.

Life story of Chamomile

Chamomile (*Matricaria recutita*)
A field of fresh chamomile is an unforgettable sight. Little golden orbs bobbing in a sea of white florets with an unforgettable smell that drifts through the air and embraces you. But it's no fun to harvest. In Egypt where we source our chamomile from the flower heads are harvested by hand by plucking off the flowers. It is a long and slow job and care needs to be taken to collect the flowers without damaging them. One picker can collect 8-10kg/day by hand with about 5000 flower heads needed to make one kilo.

Chamomile is used as a delicious tea to induce good sleep, settle restless legs and stop spasms. The exquisite yellow flowers are full of sweetness that can help to relax the nervous system. It also has a mild bitter flavour that makes it a wonderful digestive, helping to ease bloating, cramps and inflammation throughout the digestive system.

Life story of Fennel

Fennel (*Foeniculum vulgare*)
We use two types of fennel in our teas: Sweet fennel and bitter fennel. They both grow to about a metre high with long branching stems that support the most delicate leaves and 'umbrella' like yellow flowers. Sweet fennel fruits are plump and green whilst bitter fennels are dark and brown. We spent years searching for the best sweet fennel and have found it in the arid plateaus of Turkey where we get a very high essential oil content with a delicious sweetness. Our bitter fennel comes from just across the waters in Bulgaria where it flourishes as a perennial.

Fennel seed is sweet and gently warming and helps to build digestive strength by improving assimilation and preventing 'wind'. As its ascendant flower heads suggest, it spreads and moves outwards, thus preventing congestion and stagnation in the abdomen and chest. The tea is an ideal way to benefit from fennel, useful to help children's digestive colic and encourage the flow of a new mother's breastmilk.

"Rejuvenation brings you a long life, a sharp memory, intelligence, freedom from disease, youthfulness, beauty, a sweet voice, respect and brilliance." Charaka Samhita

6
The quest for quality

Health freedom

It's incredible to think that the ability to sell something as beneficial as herbs in Europe (and across the world) is becoming increasingly restricted by regulation. There is a concerted effort across Europe to increase the 'safety' of natural health solutions. There is a litany of regulations that affect us: Food Supplements Directive (2002/46/EC), Novel Foods Regulation (258.97), Nutrition and Health Claims Regulation (No.2006/1924), Human Medicinal Products Directive (2004/27/EC) and the Traditional Herbal Medicinal Products Directive (THMPD) (2004/24/EC). Of course, we need some regulation to make sure that what we use really is the correct species and outlandish claims are not made. We need to know the boundary. But we do not support restrictive legislation that limits our basic human right to health freedom: the freedom to access natural plant remedies and be informed about how they have been traditionally used.

Since we started Pukka it's been a quagmire of contradictory and, oftentimes, illogical rulings. In fact we made our very first teas using a herb called stevia. It's a remarkable sweet herb with zero calories but unbeknown to us stevia was not allowed in the EU (but was in the US and Japan). Interestingly in 2012 it has just been authorized for use. So there is hope. But sadly over the last few years other species have not fared so well; we have had to stop selling amla and tulsi in Denmark, a type of cinnamon and seaweed in Italy, bala in the USA, manjishtha in Germany and ashwagandha in lots of places. Vidanga, a herb traditionally given to children in India with no evidence of harm (historical or scientific) has been banned in the UK since 1977.

And it's getting worse. For example the Traditional Herbal Medicines Product Directive came into force on May 1 2011. It came into law the year after we set up Pukka after a seven year 'phase-in'. It's difficult to believe but the THMPD requires us to spend at least £100,000 to get a full-blown traditional medicines licence to sell a herbal product. It's technically complex, expensive, but worst of all, offers no solution to the supposed *raison d'etre* of the directive: public safety.

The herbs that are available on the UK market are not dangerous in the sense that tobacco is, or alcohol is, or even drinking too much water is. In fact herbal medicine and natural products have an incredible safety record. So much so that following an extensive review of adverse events, the chief coroner in New Zealand declared: "In so far as natural products are concerned the linkages to public safety and risk can be described legally as *De minimis no curat lex.* That is – of minimal risk importance." Which, in simple English means that in relation to the number of users, which is thought to be more than 25% of the UK population, herbs are incredibly safe. Especially when compared with the fact that pharmaceutical adverse events occur in 10% of users and are one of the leading causes of death (i.e. over 250,000 per annum in the USA). So, in this context, such a heavy-handed and restrictive law seems rather ridiculous.

As Edmund Burke said: "All that is required for evil to prevail is for good men to do nothing" and that is why we are ardent fans and supporters of The Alliance for Natural Health (www.anh-europe.org) who lobby, educate and fight for health freedom. If we want to be able to choose how we look after our health we need to stand up for our right to use natural herbal remedies.

Exemplary standards

"We are what we repeatedly do. Excellence, then, is not an act but a habit." *Aristotle*

Being truly pukka is sometimes challenging, but is always rewarding. In the early days of Pukka we realised we had to establish our own measure of excellence. Many herbs available on the 'open market' are merely 'food grade' or of variable quality. They are ok, but not the best. And we wanted the best. This means going to the source. Before we even set-up Pukka we realised that we had to develop relationships with growers who wanted to meet our standards and this meant creating standards, protocols and new economic models to help farmers qualify as Pukka growers. In pursuit of the best quality tea and products, we were led to create new ways of herb farming and a new model of business. Somehow the values we have within Pukka drive us ahead. It's as though the plants themselves give us all an enthusiasm to strive for the top and create something better, something that is beyond our current reach, but something that we can get to. And our approach to quality control sums this up perfectly.

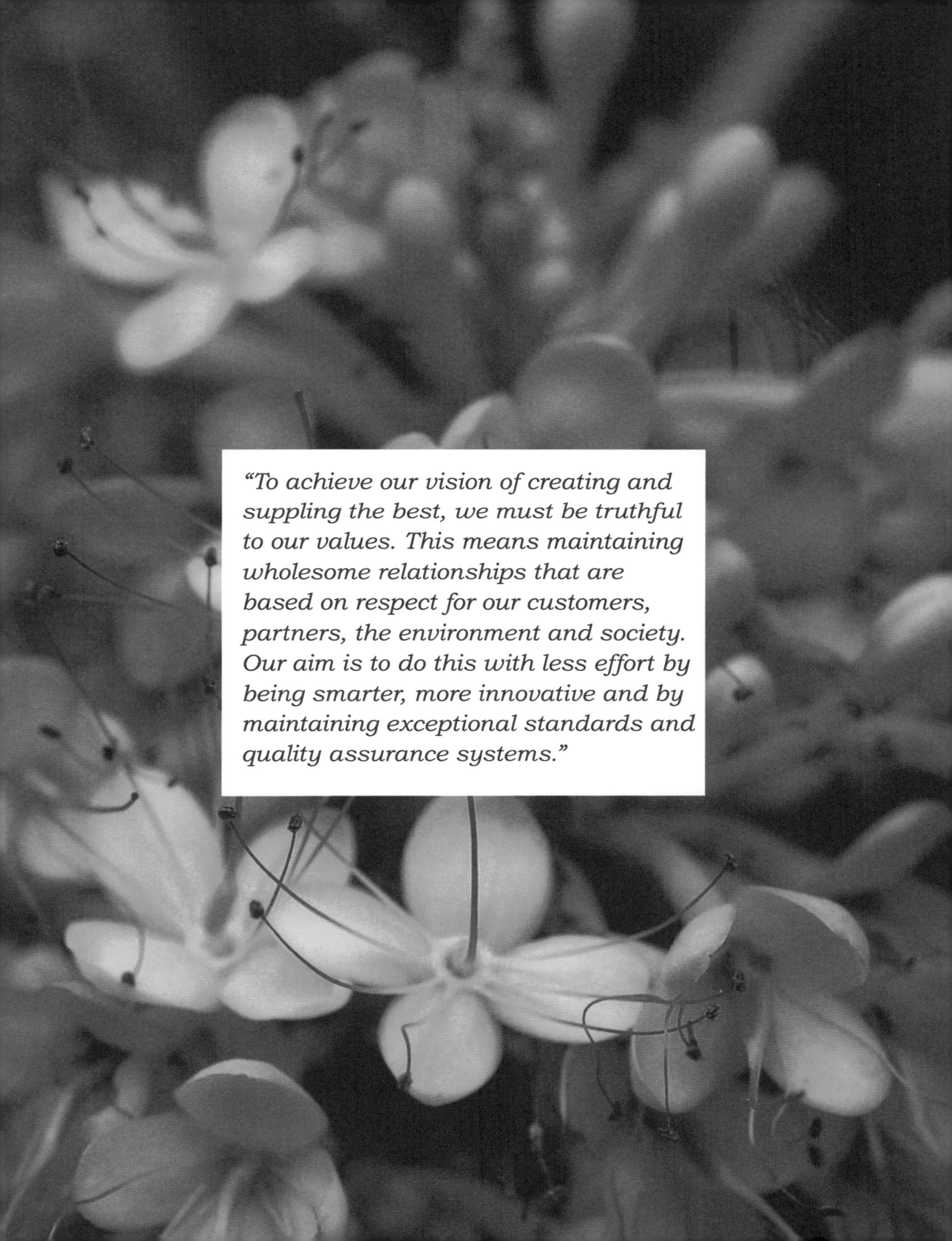

"To achieve our vision of creating and suppling the best, we must be truthful to our values. This means maintaining wholesome relationships that are based on respect for our customers, partners, the environment and society. Our aim is to do this with less effort by being smarter, more innovative and by maintaining exceptional standards and quality assurance systems."

Exemplary standards

It is important to understand that many of the farmers we work with are either poor with limited resources and/or live in very different cultural conditions to us. Just as we need help to understand the unique climactic conditions they are working in, they need help with certain aspects of growing and harvesting to the standards we require. So with Ben Heron's expert input we have developed a training programme including a training manual, a video and a booklet, to help this knowledge transfer. It's based on a specialist and pretty 'geeky' field of herbal quality standards, first established by the World Health Organisation, called Good Agriculture and Collection Practices for Medicinal Plants (or GACP for short). Its an incredible figure but 80% of the world's population depend on herbal medicine as a primary source of health care so its only right that technical experts from around the world have got together to create some detailed guidance.

The whole purpose of the GACP guidelines is to ensure that plants used as herbal medicines are grown, harvested and dried properly to optimise the quality, safety and efficacy of the finished herbal products. Another crucial aspect of GACP is to encourage the sustainable cultivation and collection of medicinal plants in ways that respect and support the conservation of medicinal plants and the environment in general.

These are really exemplary standards which have been developed because of the growing concern about the amount of medicinal plant material being produced using Bad Agricultural and Collection Practices.

The growing demand for herbal products has also led to over-harvesting from the wild, causing concern over the long-term environmental impact on the availability of certain herb species if they are not collected in a responsible manner.

This call for greater quality assurance highlights the need to provide training to herb growers, collectors and processors which is something we relish at Pukka. We provide ongoing training to farmers in these following key areas:

- **Hygiene and cleanliness** to promote quality and prevent microbial contamination.
- **Plant identification** to ensure that the herbs they are growing are the correct species and that we have identity checks in place. This means working with University botanical departments and pharmacogsonists (specialists in medicinal plant identification).
- **Optimising efficacy** through growing the plants in a suitable environment, in appropriate soil conditions and ensuring that they are harvested at the correct time.
- **Increasing the yield** of the crop through using the best organic farming practices so that the farmers get the best return for all their hard toil.
- **Sustainability** to ensure that the plants that are collected in a manner that allows the plant population to regenerate year after year.
- **Documentation and traceability** to ensure that each batch of herbs we buy can be traced back to their origin. This is something we are obsessed about so that we know where each species comes from and how it has been grown, harvested and dried.

GACP encourages our supply chain to work together to prevent problems. Just as optimum health is created by the harmonious functioning of mind, body and soul, so effective herbal remedies are created by the combined knowledge and effort of the farmer or collector, manufacturer and practitioner.

After the harvest back in the lab

Of course, our efforts to bring you the best quality herbs doesn't stop on the farm. We have a highly skilled team of quality experts and scientists ensuring that the plants we buy are really Pukka. We like to blend ancient wisdom with modern insights using both time-honoured as well as contemporary analysis techniques.

The primary traditional method to interpret the quality of the herb is to use something called 'organoleptic' assessment which means using our senses to check the colour, taste, friability and the overall vibrance of a herb. And then we do all sorts of interesting technical analysis at our lab in the UK such as: 'swell index' analysis to check how much mucilage is in herbs such as marshmallow root, and essential oil extraction to check the level of aromatic volatiles in herbs such as lavender. We also do something called "High Performance Thin Layer Chromatography' (HPTLC) which is a way of showing what phytochemicals are in the herbs we use in our products. The concentration on the 'bands' reveals the quality of the batch.

Here is an HPTLC Chromatograph showing the consistency of six Andrographis panniculata extracts grown on a project we work with in South India. The blue band shows the presence of a compound called 'andrographolide' and the density of the band shows how much is present. Through implementing more efficient organic cultivation techniques and better post-harvest drying and grading we have managed to get the andrographolide content up from 1% (the minimum standard) to 6%.

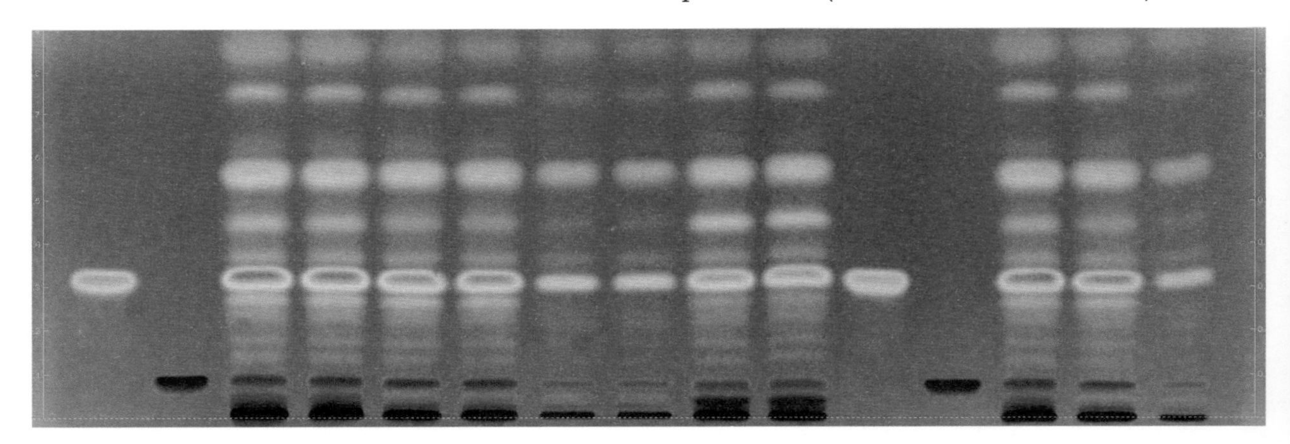

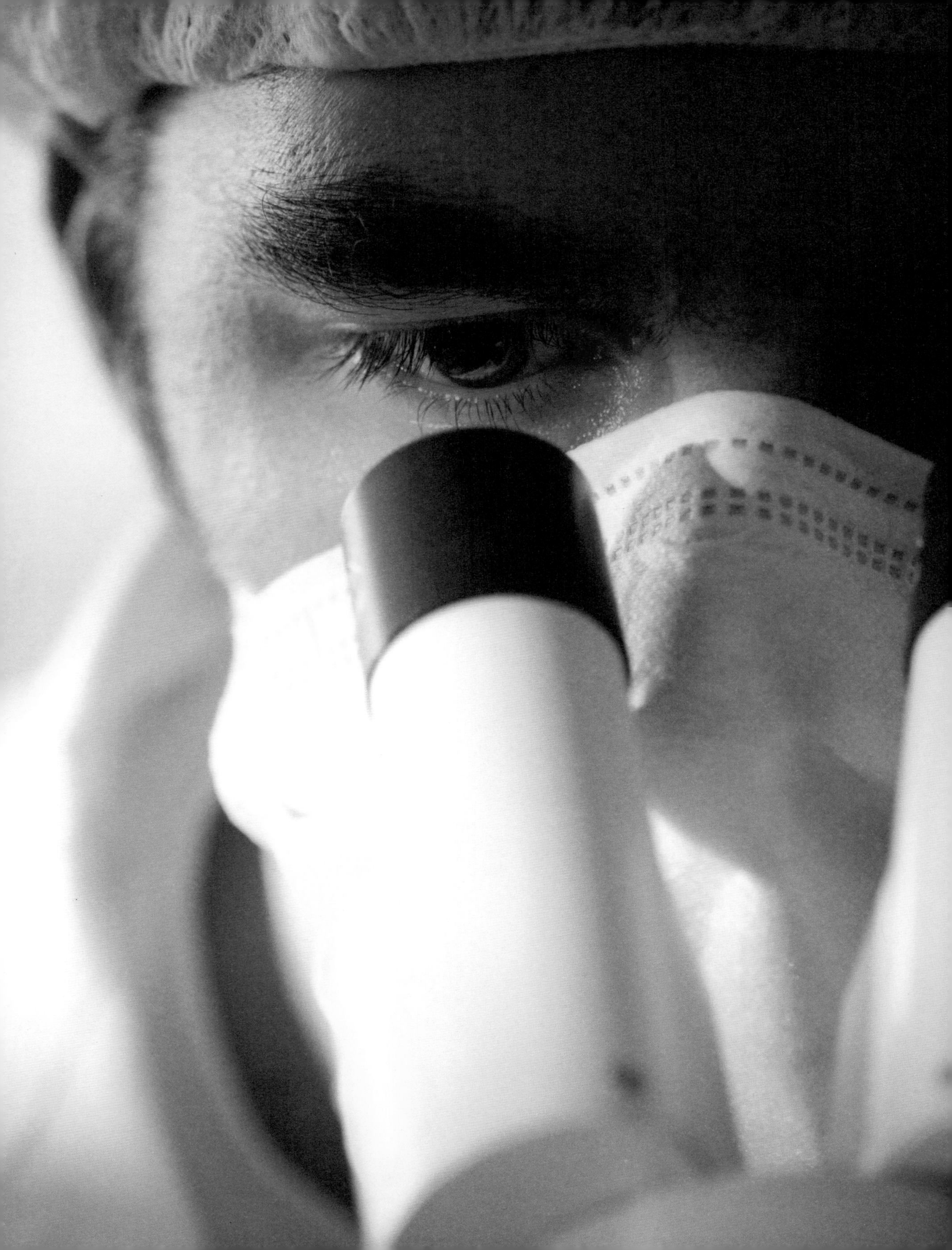

Herbal sustainability

It is remarkable but perhaps not surprising to think that most herbs are harvested from the wild. Have you ever thought where your last cup of elderflower, licorice or limeflowers came from? These herbs are rarely cultivated, as they are so readily available for free in the local environment. This means that herb collectors go out into the forests to pick what they need. Herb collectors in the countries where many Ayurvedic herbs are grown are usually the poorest of the poor. They often do not own land and are dependent on annual wild herb harvests to supplement their income. However, such unregulated harvesting is putting the sustainability of herbs – and, by extension, herbal medicines – into question.

In January 2004, Alan Hamilton, a plant specialist working with the World Wildlife Fund, released a paper on the threat to the herbal community from the indiscriminate over-harvesting of medicinal herbs. He noted that approximately 75 per cent of all medicinal herbs come from the wild: there are 50,000 species that are used as medicines and 10,000 of these are threatened. This means that a staggering twenty per cent of all herbal species used throughout the world are under threat! For example, licorice, goldenseal, devil's claw, guggul, sandalwood and slippery elm are all threatened in their natural habitats.

"Uncontrolled wild harvesting is threatening the medicinal plants on which the herbal medicines industry depends." Alan Hamilton, World Wildlife Fund

An organic business deal.

The threat to herbal supplies

It is estimated that the Ayurvedic pharmacopoeia includes upwards of 1250 species, with approximately 300 of these in regular demand. Similar figures exist for Chinese and western herbal medicine. In India and Sri Lanka most herbs come from the wild: in excess of 90 per cent of herbal material used in Ayurveda comes from the forests, mountains and plains of the Indian subcontinent and is sourced in an unregulated manner. In other parts of the world there is similar pressure, with 80 per cent of species coming from the wild in China and up to 99 per cent in Africa. Global demand for herbal medicine has increased by an estimated ten to twenty per cent per annum in the last decade and so the pressures to over-harvest are immense.

However, the harvesting of wild herbs is a relatively accessible source of income for people without land or a regular job. Herbs grow for free in the wild and trading collected wild plants can support a family. In the higher altitude region of Nepal all families harvest herbs and it can account for 15–30 per cent of their income.

The escalating demand for herbal medicine coupled with the needs of low-income families puts increasing pressure on natural habitats. Having been involved in trying to grow jatamansi and kutki in the Himalayas, we have seen these pressures first hand, as the local pickers trot down the hillsides laden with unsustainably harvested wild herbs. Instead of waiting until the right time to start wild collection, it has now become a race against time to get to the herbs before other collectors, and it is common to see people collecting some species three months before the traditional time which means that the remaining population does not even have the chance to flower and spread their seeds. With no hope of regeneration, the precious jewels of the land are being stripped.

The future

The majority of herbal species are not on the verge of extinction but many are threatened. As supporters of the herbal community and bearers of its heritage; we must make sure that we act in a truly Ayurvedic fashion and help to prevent 'disease' before there is a problem. Therefore we source herbs, oils and foods that aim to conserve the integrity of nature without damaging it. In practice this means that we always buy organically certified and sustainably harvested ingredients. Its really very simple: protecting nature protects us individually as well as promoting global health.

7

Connecting people & plants

Connecting people and plants

"The essence of all beings is Earth. The essence of the Earth is Water. The essence of Water is plants. The essence of plants is people..." Chandogya Upanishad

There are some exemplary schemes that have been developed over the last few years to help people and plants connect. By 'connect' we mean to join them together in such a way that the interwoven nature of their existence is appreciated and respected. It is a bond that is both inspiring and indelible. Here are a couple of schemes that we are fortunate enough to have played a small part in:

Fairtrade projects
We are privileged enough to work with certified Fairtrade projects in India, China, Vietnam, Ghana and Egypt. Fairtrade is what it says. It's a system of certification that ensures that the people doing the hard graft of growing the produce are paid a fair wage and treated fairly thus benefitting the community as a whole. There is usually a guaranteed minimum price as well as a Fairtrade premium paid to the producers to help social and economic investment and improve their livelihoods. It's another example of intelligent capitalism. The growers get a better deal, the consumer can be content that their money is helping to improve the lives of people less privileged than themselves and the business involved can be satisfied that they are contributing to social welfare of their partners.

Whilst many of the herbs that we buy are not available Fairtrade certified it is still very important to us to ensure that the social welfare of the farmers we work with is considered in the 'whole' of our sourcing practices. This means developing a long-term relationship, working with them to get consistently high quality and making sure that they get a good deal.

Fairwild
The Fairwild certification is a recent scheme started to ensure sustainable wild collection of herbs. It is an example of incentive based conservation. It combines the principles of FairTrade, to ensure collectors are fairly treated and appropriately paid, with the practice of sustainable wild collection. Fairwild ensures the long-term survival of the species and the community. It ensures that the whole is considered; people, the plants and the eco-system are all equally important. In this way FairWild certification can transform potentially destructive wild-harvesting practices into a powerful tool for conservation.

We are working with Fairtrade projects in India, Khazaksthan, Bosnia and Albania.

Our environmental commitment

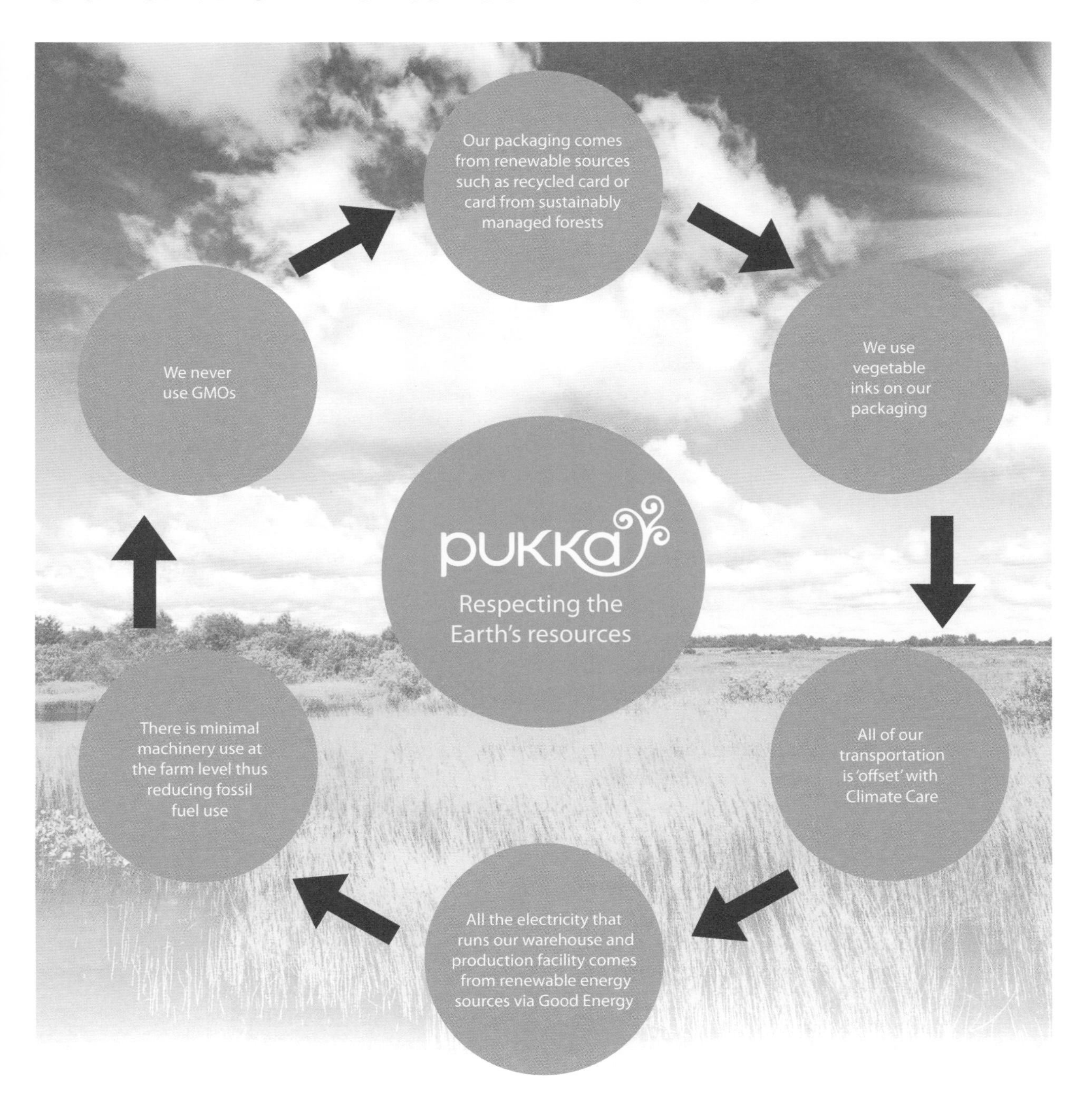

Since 2005 we have offset our carbon emissions with Climate Care to ensure that all the carbon emissions generated by our moving our goods around the world are offset.

Whilst we recognise that this is not an absolute solution it is an indication of our intention to work in a manner that acknowledges its impact on the environment.

The Hornbill project: Hope for the future

Our hornbill project is in its early developmental stage. This is a project that is really the essence of what Pukka is about: people, plants and habitats. The great pied hornbill nests in the giant trees of the Western Ghats that stretch along the west coast of India. As this habitat becomes increasingly under pressure the home of the hornbill is also compromised. The presence of the hornbill has become a symbol of the health of the ecosystem and hope for the future. And we want to support it.

One of the fruits we want to harvest from this area with Fairwild certification is from the giant Bibhitaki trees: Bibhitaki (*Terminalia belerica*) is one of the fruits in the famous Triphala formula, but its wood is also sought after as an easy source of income. One of the problems is that the great hornbill nests in the Bibhitaki tree. So our plan is to establish a project that incentivises collectors to gather the Bibhitaki fruits and earn an income that will stop them from needing to chop down the trees to earn money and, in this way, protect the home of the hornbill. Watch this space.

Kutki in the Himalayas: Trials and tribulations

Cultivation is one approach to conserving threatened species such as Kutki and is of critical importance if the species is wiped out in the wild. We have been involved in a project with Biolaya to grow this in accordance with CITES regulations (CITES is the international convention that oversees trade in endangered species). However due to the frustrations of Indian bureaucracy it has taken five years to produce and successfully export the correctly grown crop and aged one member of our team ten years in the process.

Guggul in Rajasthan

Guggul, or Indian Myrrh, is a shrub that grows in arid regions, primarily in Rajasthan and Eastern Pakistan and is used for clearing toxins and helping metabolism. It is harvested by cutting its trunk or stem with a knife, which releases a milky coloured resin. The plant is left for a few days for the resin to ooze out of the wound, coagulate and dry in the sun. This dried resin is collected and sold as Guggul. As with most threatened species it is under threat because demand has become greater than supply. Each plant can only produce relatively small quantities, and if collectors try to extract too much (which seems to be the norm) then there are high chances the plant will dry up and die.

Cultivation of Guggul is quite a brave undertaking. It takes 6-8 years before the plant can be harvested, after which each plant produces approximately 100g resin per year. The plants are sensitive and there are chances that some of them will die from infections etc. Because of this there are very few people cultivating Guggul and even fewer who are certified organic. So we count ourselves as extremely fortunate to be working with a totally committed family in Rajasthan that have made this project a success.

WWF: Inspiring positive change

We have teamed up with WWF-UK to promote awareness of the importance of the connection between people and nature. WWF is a global leader in conservation and we are working together to raise awareness of the importance of sustainability for a future in which people and nature can thrive.
As WWF are founding partners in establishing the FairWild standard for harvesting wild herbs, we have created a new FairWild tea, Peppermint & Licorice, to celebrate the benefit that plants can bring to people and nature. Our work together is all about the positive change that incentivised conservation can have on our health. It's about spreading the message that through a cup of herbal tea we can all be connected and that working together for mutual benefit really brings positive results.

20p
donated to
WWF
with every
pack
pukka
peppermint
& licorice
deliciously refreshing
organic herbal tea
WWF
PUKKA
SUPPORTS WWF'S
WORK TO
HELP PEOPLE
AND NATURE
THRIVE
Net Wt 30g e (1.05 oz)
20 herbal tea sachets

Bringing wild species into cultivation

With the exception of certain species which need to be collected from the wild or can easily be sustainably harvested, we are continuously working to bring species into cultivation. During the last few years we have been successful in cultivating species such as Brahmi, Gotu Kola, Andrographis and Shankhapushpi, which are traditionally collected from the wild. In the absence of controlled collection systems, such as those implemented through FairWild, bringing plants into cultivation is an important conservation measure.

Planting 108 tree species in Karnataka

For the last four years we have been donating towards a tree planting project in the forest around Hosagunda (the organic farm owned by CMN Shastry). The aim of the project has been to plant 108 species of rare and endangered trees every year for five years.

Alongside the 108 Tree Species project CMN Shastry has also been doing a rainwater harvesting project, which involves digging hundreds of trenches in the forest to capture rainwater during the monsoon to re-charge the water table. This has had a dramatic impact on the availability of water in streams that were once perennial but have dried up in recent years, allowing farmers to grow two crops in a year rather than one.

1
LOVE
A pukka life

Thanks & Acknowledgements

When we started to write this book we wrote to our customers, suppliers and friends and asked them to send in quotes and pictures that reflected on why they liked working with Pukka. We didn't realise that so many would reply! As we couldn't include them all, here is a huge Pukka cup of 'thank yous' to everyone who has supported us over the years and helped our dream come true. We also asked our staff to send in quotes, recipes and comments about why they like working at Pukka. Sorry we can't include them all but here are some of our favourites:

"If you are in control you are not driving fast enough."
Emerson Fittipaldi via Timothy Westwell

"Dance like no one is watching, Love like you'll never be hurt, Sing like no one is listening, Live like it's heaven on earth."
William Pugh via Amanda Heron,
Business Development Manager

"Be who you are and say what you feel because those who mind don't matter and those who matter don't mind."
Dr Seuss via Ben Heron,
International Supply Manager

"I must create a system or be enslaved by another man's; I will not reason and compare: my business is to create."
William Blake via Suze Pole,
Creative Inspiration

"After one look at this planet any visitor from outer space would say 'I want to see the manager'." William S Burroughs via Barry Moore, Quality Manager

"Fail to plan, plan to fail." Winston Churchill via Ben Whitbread, Warehouse Manager

"Always look on the bright side of life."
Monty Python via Donna Best,
Customer Service Manager

"Every day do something that will inch you closer to a better tomorrow."
Doug Firebaugh via Graham Collins,
Operations Manager

"Football's not a matter of life and death ... it's more important than that."
Bill Shankley via Nick Britton and the warehouse team

"Being deeply loved by someone gives you strength, while loving someone deeply gives you courage."
Lao Tzu via Shamini Singh,
Sales Team Manager

"Try not to become a man of success, but rather try to become a man of value."
Albert Einstein via Marin Anastasov,
Sourcing Manager

Q: "Can we really afford to Tim?"
A: "Well, can we afford not to?"
Conversation between Sebastian and Tim.

"Harnessing the incredible Health Benefits of Organic Herbs." Pukka Herbs

A big thanks to all of our photographers: Louise Farmer, Ben Heron and Sebastian Pole.

8
Pukka people

How we get there:

Inspiring staff – the people behind Pukka

So we had this big idea but ideas are ten-to-the-dozen, it's making them happen that counts. From the outset we needed some exemplary people to help bring the Pukka vision to life. It has been the most challenging aspect of creating a business and we have had to learn lots about building a team but, fortunately, we have some incredible people working with us. Great companies are created by the people that work for them. It begins with their dreams and how they see the world. It takes shape in their ambitions and their day-to-day actions. Inspired companies are created by people who want to make a difference and who play the long game by giving their small actions big purpose. They can do this because their values give them clarity, patience and strength. Pukka's future is always the sum total of its employees' dreams and ambitions. As with any community or tribe, reflecting on our shared values will help us realise these dreams together.

Our staff are the heart of Pukka. They make it possible to smoothly run what is a very complex and challenging business. They work in very diverse areas: from farmers in remote agricultural parts of the world, to complex European legislation, from technical herbal quality analysis, to the intricacies of web management, from global financial institutions to our very sophisticated customers. We aspire to make Pukka a platform for our staff to develop and grow. People are attracted to work with Pukka because they share the same values and concerns as Pukka – healthy people and a healthy planet. Shared values create a synergy and a lot of energy.

To be the best we need the best and Pukka is made of many talented, educated, well-informed and committed people. We have scientists and eccentrics (sometimes they are one and the same), artists and herbalists, farmers and gardeners, customer carers, a warehouse full of hard-working fellows who keep everything coming and going efficiently and on time, managers and leaders, financial wizards and so many more who help the company achieve its vision of bringing people and plants together. Different as everyone is, it's our Pukka values that motivate them and make them part of our family circle of trust.

pukka

Our family tree

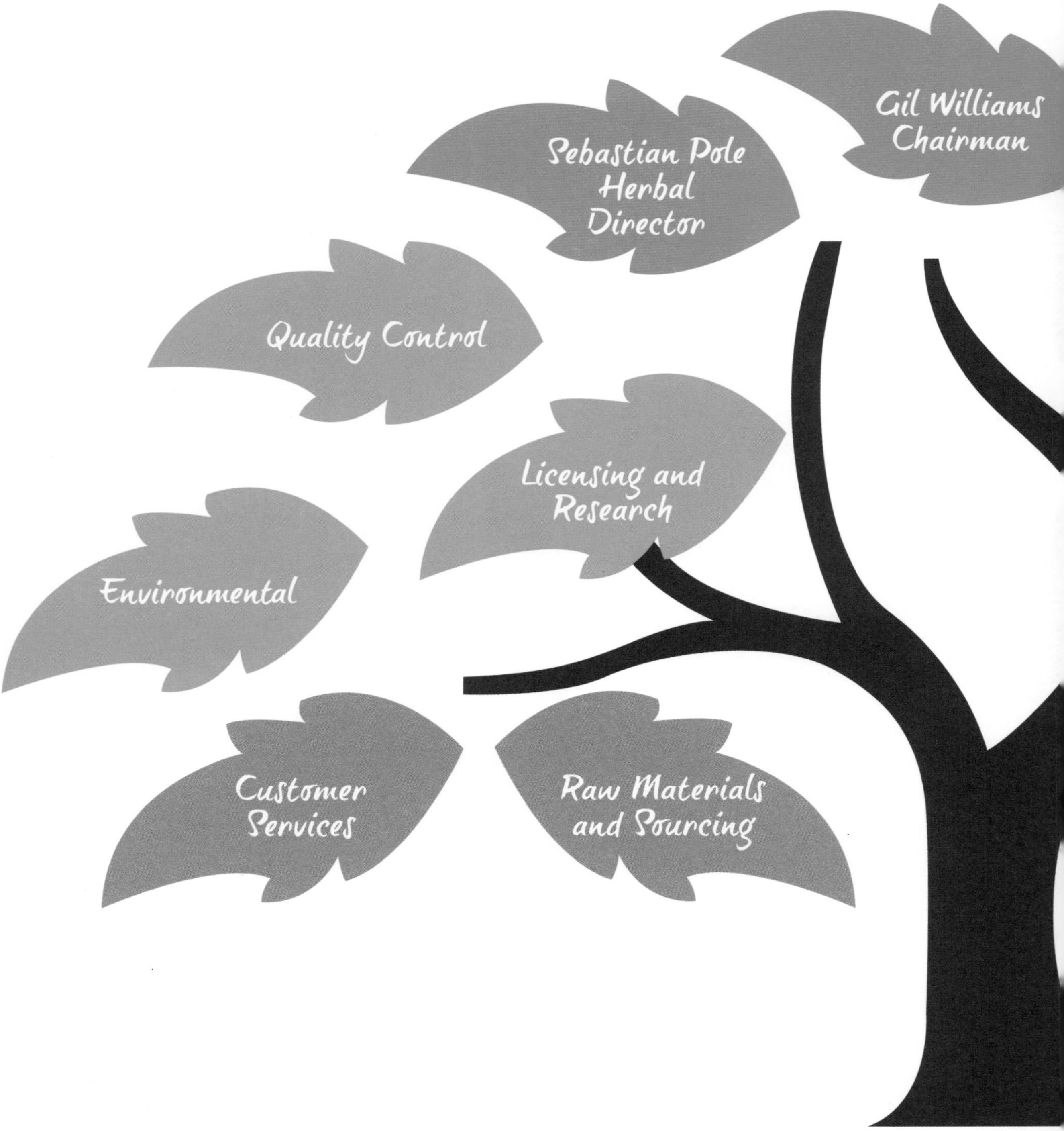

Gil Williams
Chairman

Sebastian Pole
Herbal
Director

Quality Control

Licensing and
Research

Environmental

Customer
Services

Raw Materials
and Sourcing

Tim Westwell
Managing
Director
International
Sales
UK Sales
Marketing
Finance
HR and IT
Purchasing
Operations

Pukka People

Ali Wilde, Head of Marketing

"What can I say about Pukka? It's a phenomenal company to work for. If you've ever felt disheartened by the commercial greed of some businesses, then a spell here will demonstrate just how the infectious passion of two guys can really make a difference to the lives of others. Their philosophy of ensuring people benefit from the natural power of plants works: from the way they work on a one to one basis with our farmers, to the health benefits of our consumers who use our herbal products. Every day I wake up happy about going to work, sure it's fast paced and sometimes I get home wondering where the day has gone, but that is what life is about - diversity and living it to the full, why on earth would you drag yourself to a place you don't want to be?"

Rhiannon Philips, Head of Finance

"The thing that really enthuses me about working at Pukka is the potential to help build an ethical and exciting business across the world. I enjoyed the formative years of my career in large corporate companies and I thank them for my personal development, however what really motivates me is the sense of satisfaction I get from being able to add value to a business that is really making a difference. Yes there are challenges (mainly in managing our growth!) as with any business, but at Pukka we have a strong team of talented and committed people whose hearts are in the job as much as their heads."

Donna Best, Customer Services Manager

"One of Ben Heron's, our International Sourcing Manager's presentations about the sourcing and growing of our herbs in India and the impact it has on the people and environment involved really brought the whole story of Pukka and the herbs we sell back to life for me – it is so easy otherwise (in my job anyway) for them to just become codes and numbers. I love being part of a young company and being able to contribute to processes, procedures and aptitudes that I hope will endure."

Kay Collins, Packaging & Procurement Manager

"There's a certain excitement in being involved in such a progressive company, of building and working in a team that faces fresh and demanding challenges each day. I hope that I have played a small part in solving those challenges and supporting the rapid growth that Pukka has achieved and "made that difference" that I desired to do when I first started."

Ben Heron, International Supply Manager

"I consider myself to be extremely fortunate to work with Pukka's suppliers, spending time with the farmers and the plants. This gives me a unique perspective from which I get to see the enormous amount of care and effort that goes on behind the scenes to produce excellent quality herbs, as well as the huge number of people who benefit from growing them. On a personal level what I love most about working at Pukka is that, whilst being a successful business firmly rooted in the realities of commerce, its motivation is entirely heartfelt, driven by the commitment of a group of people who are passionate about people and plants, and genuinely wish to make the world a better place."

Amanda Heron, Business Development Manager

"People, passion and plants paving the way to a better future for all is instrumental in everything I do, working in an environment where values and ethics are core to making a difference in the world is what drew me into a Pukka Life!"

Some of our awards

great
taste
gold 2010

Some of our rewards

Life started to form on Earth 3.5 billion years ago
In the last ten years we've made over
1 billion
cups of tea
To count to one billion would take about 100 years

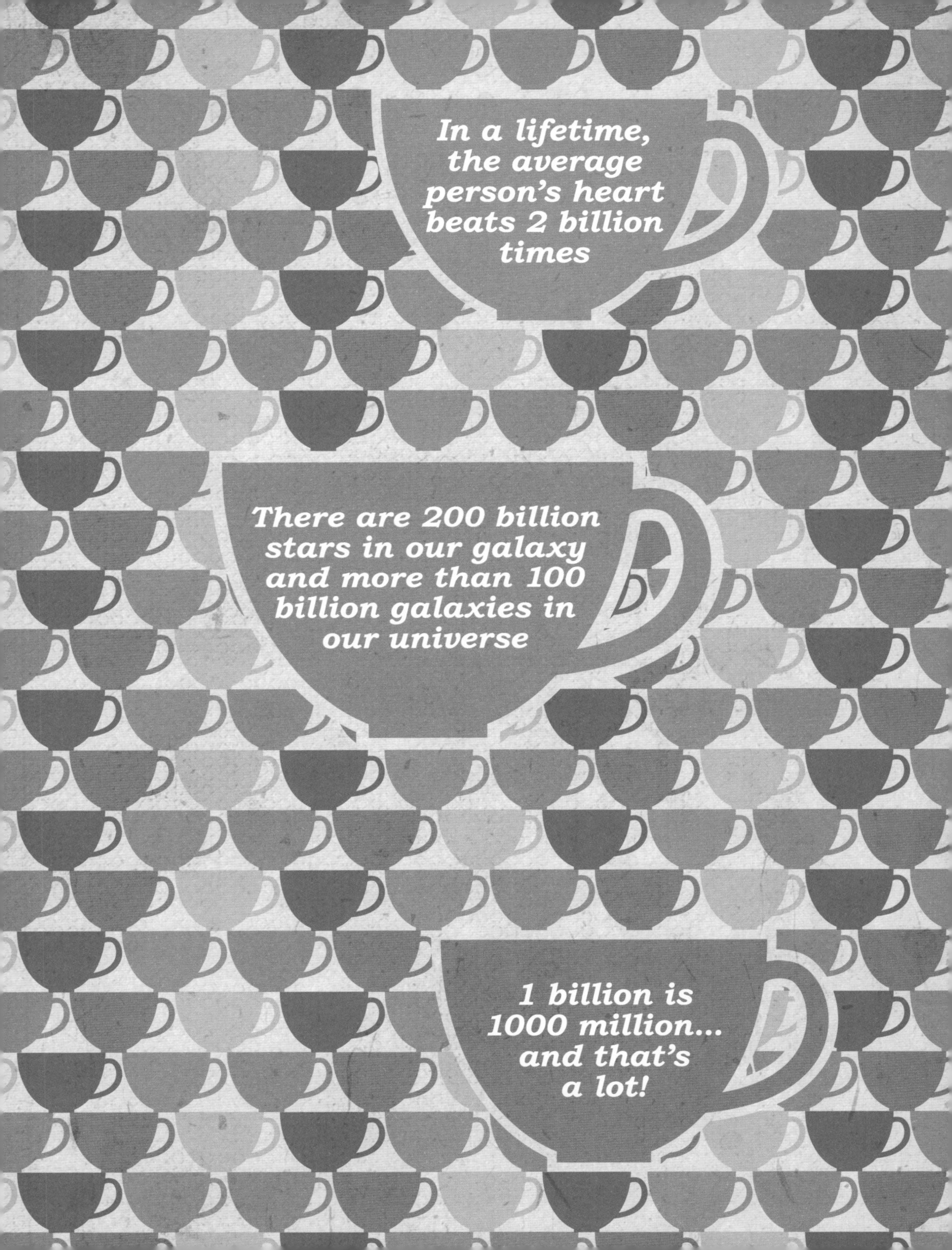
In a lifetime, the average person's heart beats 2 billion times
There are 200 billion stars in our galaxy and more than 100 billion galaxies in our universe
1 billion is 1000 million... and that's a lot!

Expect wonderful things

When we started Pukka Herbs we knew that we were going to create something special: we were both utterly passionate about bringing plants and people together and we set off with the inspiring tailwind of Ayurveda behind us. We really wanted to be a part of the positive change that was at the heart of the organic health community and so we dived in, head first. And, wow, what a welcome we have received!

Today we are making millions of teas and capsules, growing acres of herbs and introducing thousands of people to the wonders of traditional herbal wisdom. There are now 50 of us working together at Pukka. We are a pretty eclectic, often hectic, but always energetic and inspiring bunch of people. We're making the best of British and do 99% of our manufacturing in the UK from organic ingredients sourced from many friends around the world. In our own small way we hope we're helping people to live healthier, happier lives and to fulfil their potential on this beautiful planet that we all share together.

So, what next? We are even more excited about Pukka Herbs now than when we started ten years ago. So many people have helped us realise this incredible potential and we are going to do our utmost to manifest it. We are going to carry on our mission to help our customers, partners and staff improve their lives through sharing the amazing power of herbs. There are plenty of seeds yet to be sown and a lot of wonderful things still to be shared. Here's to you.

With our best wishes

Sebastian and Tim